The Analysis of Risk Management of community water projects in Sub - Saharan Africa

A Case Study of South Sudan

Tamburo M. Renzi, Ph.D

authorHOUSE

AuthorHouse™
1663 Liberty Drive
Bloomington, IN 47403
www.authorhouse.com
Phone: 833-262-8899

© *2023 Tamburo M. Renzi, Ph.D. All rights reserved.*

No part of this book may be reproduced, stored in a retrieval system, or transmitted by any means without the written permission of the author.

Published by AuthorHouse 09/28/2023

ISBN: 979-8-8230-0022-2 (sc)
ISBN: 979-8-8230-0020-8 (hc)
ISBN: 979-8-8230-0021-5 (e)

Library of Congress Control Number: 2023901728

Print information available on the last page.

Any people depicted in stock imagery provided by Getty Images are models, and such images are being used for illustrative purposes only.
Certain stock imagery © Getty Images.

This book is printed on acid-free paper.

Because of the dynamic nature of the Internet, any web addresses or links contained in this book may have changed since publication and may no longer be valid. The views expressed in this work are solely those of the author and do not necessarily reflect the views of the publisher, and the publisher hereby disclaims any responsibility for them.

Contents

Affidavit xvii

Abstract xix

Introduction 1
   Research Background 1
   Problem Statement 5
   Purpose of the Research 6
   Research Aim and Objectives 7
   Research Questions 7
   Significance of the Research 8
   The Constraint of the Research 10
   Assumptions of the Research 12

Literature Review 14
   Introduction 14
   Water Systems and the Origin of Water Problems in South Sudan 14
   Water Policies in South Sudan 18
   Reasons for Implementing Water Kiosks as Community-Based Projects ... 23
   The Essence of Water Stands in South Sudan 28
   Case Example of Previous Public Water Projects 30
   Other Projects 34
      Lesson Learned from the Ghanaian Public Water Scheme 35
      Internal Community Support 35
      The Role of External Support 37

Water Issues and the State of Affairs in South Sudan 42
Conceptualizing Risk and Risk Management 44
Processes of Risk Management..45
    Risk Identification 45
    Risk Assessment 46
    Risk Response Planning 46
    Monitoring and Controlling...47
    Identifying Risk Factors 47
    General Characteristics of Samples................................... 48
Water Politics in the Nile Basin Region 53
Water Systems and Water Stress in the Nile Basin Region......54
Traditional vs. New Models for Water Supply 58
    The Inefficiency of Public Water Initiatives and the
    Need for Water Tariffs 58
    Decentralization in Water Supply Systems and
    Community-Based Water Management (CBM) 60
    Evaluating the Sustainability of Community-Based
    Water Initiatives 64
    Increasing the Sustainability of Water Initiatives..................... 68
    Factors Reducing Sustainability of Community Water
    Initiatives 75
Town Overviews 78
About Tambura County and the People 78
Tambura County Water Supply Projects 80
Tambura West Water Supply Project.......................................81
    Duties of Water Management Association (WMA) 84
    The Water Usage and Mode of Payment 84

Financial Management in the Water Sector 86
Mamenze Water Supply Project 88
About Tambura East Water Project 89
Establishment of the Project 90
About Yambio County and Its People 91
About Yambio Community Water Supply 91
Kpirabe Community Spring Water Project 92
    Yabongo-Napare Water Project 93
    Ikpiro Water Project 94
    About Juba County and the People 95
Juba Community Water Supply 96
Gudele West Water Project 97
Gumbo Water Treatment Plant 97
Causes of Water Shortage in South Sudan 98
    River Flow and Rainfall 98
    Return of Refugees 99
    Local and International Bodies 99
    USAID's Findings 100
    Findings of African Development Bank (AfDB) 101
    The Significance of Community-Based Water Kiosk Projects 102
Risk Management 103
    Risk Identification 104
    Risk Assessment 105
    Risk Response Planning 105
    Monitoring and Controlling 106
Operations Management 106

- Cost Management 108
  - Resource Planning 108
  - Cost Estimating 108
  - Cost Budgeting 109
  - Cost Control 109
- Entrepreneurial Actions and Innovations 110
- Summary of Literature Review and Gap 113

Research Methodology 115

- Introduction 115
- Research Design 115
- Research Approach and Methodology 116
  - Research Design Option 117
  - Research Philosophy: Interpretivism 118
  - Research Approach 119
  - Implication of Lean Six Sigma Approach 119
  - Areas Requiring Changes in Water Kiosk Projects Under Lean Six Sigma Approach 120
  - Methodological Choice: Simple Mixed Method 121
  - Strategy: Action Research 121
  - Process of Action Research 122
  - Time Zone: Longitudinal 123
- Techniques and Procedures 124
  - Research Methodology 124
  - Data Analysis 124
  - Identified Primary Stakeholders 125
  - Identification of Flaws and Waste in the Water Projects ... 126
- Targeted People 127

Sampling Method 127
Data Collection Instrument 128
Data Analysis Methods 129
Validity and Reliability 131
Instruments of Validity 132
Data Collection Process 132
Limitations of the Research Study 133
Research Discussion, Analysis, and Findings 135
Introduction 135
Data Management and Representation 136
Findings 136
    Findings from the Water Managers 136
    Findings from the Water Organizations 166
    Findings from the Water Vendors 178
Summary of Findings from the Interviews 201
    Challenges of Water Projects 201
    Reasons Why the Water Project Will Not Live Longer 203
    Metering and Payment Methods 205
    Safety of Water for Drinking 206
    Security of the Water Projects 208
    Risks of Managing Community Water Kiosks 209
Conclusion and Recommendations 211
Introduction 211
Discussion of the Findings 214
Similarities and Differences with Other Studies 215
Importance and Implications of the Findings 218

 Researcher's Opinion on the Findings 219

 Contribution to Knowledge 220

 Contribution to Practice 221

 Limitations of the Research 221

 Research Objectives 222

 Future Research 222

 Recommendations 223

 Summary 228

List of Abbreviations 229

References 231

List of Appendices 247

 Appendix 1: Summary of Interviews Conducted 247

 Appendix 2: Letter of Introduction from LIGS University 249

 Appendix 3: List of Questionnaires 250

LIGS University USA 257

 Research Questionnaires for Water Vendor 257

List of Figures

Figure 1. Community-Based Water Management Unit..................25
Figure 2. Ages of Respondents 49
Figure 3. Population Structure of Respondents.............................51
Figure 4. Peak Hours of Water Operations 52
Figure 5. Life Cycle Analysis..66
Figure 6. Broader Institutional Arrangement of Water
  Supply and Management 82
Figure 7. Structure within Water Management Association
  (WMA)...83
Figure 8. Sequential Flow of Financial Management....................88
Figure 9. Kpirabe Water Source...93
Figure 10. Ikpiro Water Source..94
Figure 11. Saunders Research Onion 116
Figure 12. Research Design Options... 117
Figure 13. Respondents in the Water Manager Questionnaire..... 137
Figure 14. Occupation of the Respondents................................... 138
Figure 15. Education Level of Respondents................................. 139
Figure 16. Experience Level of Respondents 140
Figure 17. Respondents' Age Group.. 141
Figure 18. Number of Projects and Duration 142
Figure 19. Planning and Implementation of the Projects 142
Figure 20. Reasons Behind Unsuccessful Project
  Implementation.. 143

Figure 21. Stage of Projects Posing Challenges to Water Managers 145
Figure 22. Challenges in Project Management 146
Figure 23. Legal Risks 147
Figure 24. Political Risks 148
Figure 25. Financial Risks... 148
Figure 26. Technical Risks 149
Figure 27. Security Risks 150
Figure 28. Management Risks 151
Figure 29. Solutions to Reduce Risks.. 152
Figure 30. Presence of Community Guidelines/Policies.............. 153
Figure 31. Implementation of Guidelines/Policies...................... 154
Figure 32. Still Working After Ten Years?.................................. 155
Figure 33. Reasons for Project Shutdown 156
Figure 34. Project Sustainability—Harmonized Policies............. 156
Figure 35. Project Sustainability—Capacity Training 157
Figure 36. Project Sustainability—Water Rates.......................... 158
Figure 37. Project Sustainability—Customer Control................. 158
Figure 38. Project Sustainability—Regular Maintenance............ 159
Figure 39. Project Sustainability—Control Fetching Time.......... 160
Figure 40. Project Sustainability—Periodic Monitoring 161
Figure 41. Improvement of Water Projects—Stakeholder's Involvement 162
Figure 42. Improvement of Water Projects—Regular Meetings 163
Figure 43. Improvement of Water Projects—Timely Supervision 164

Figure 44. Improvement of Water Projects—Alternative
    Funding 165
Figure 45. Improvement of Water Projects—Others 166
Figure 46. Water Organizations 167
Figure 47. Years of Operation 168
Figure 48. Number of Employees 169
Figure 49. Number of Operations in Progress 170
Figure 50. Previously Faced Risks 171
Figure 51. Recommendations for Risk Handling 172
Figure 52. Involvement of the Community 173
Figure 53. Engaging the Community 174
Figure 54. Water Projects Facing Challenges 175
Figure 55. Challenges of Water Projects 176
Figure 56. Planning to Stop Risks From Reoccurring 177
Figure 57. Measures to Stop Risks From Reoccurring 178
Figure 58. Gender (Water Vendors) 179
Figure 59. Education Level 180
Figure 60. Years of Experience 181
Figure 61. Age Group 182
Figure 62. Residents or Not 183
Figure 63. Number of People in Household 184
Figure 64. Source of Water 185
Figure 65. Number of Years Water Project Has Served 186
Figure 66. Who Fetches The Water—Mother 186
Figure 67. Who Fetches The Water—Father 187

Figure 68. Who Fetches The Water—Female Child 188

Figure 69. Who Fetches The Water—Male Children 188

Figure 70. Overwhelmed with Work 189

Figure 71. Time of Day and Feeling Overwhelmed 190

Figure 72. Is the Water Safe for Drinking? 191

Figure 73. Unsafe Drinking Water (Non-Treated) 192

Figure 74. Unsafe Drinking Water Contaminated 193

Figure 75. Reasons for Unsafe Water 194

Figure 76. Usage of Fetched Water 195

Figure 77. Importance of Water Kiosk Projects 196

Figure 78. Lack of Capacity 197

Figure 79. Lack of Regulations 197

Figure 80. Reluctance in Payment of Bills 198

Figure 81. High Demand 199

Figure 82. Water Project Under Risk 200

List of Tables

Table 1. Sustainability Evaluation Model 65

Table 2. Five Sustainability Indicators of Community Water
           Initiative 67

Table 3. Eight-Stage Framework for Maximizing
           Sustainability of Rural Water Initiatives 71

Affidavit

By inserting a seminar paper into the Learning Management System of LIGS University, I, Tamburo Michael Renzi, an Interactive Online PhD student, honestly declare that I have prepared this dissertation thesis myself with the help of my lecturer and using only the literature presented in the paper. I further confirm that I have no objection to the lending or publication of this dissertation thesis or part thereof with the approval of LIGS University.

Abstract

Risk management is one of the important planning processes in any given project. It involves identifying, analyzing, and responding to any risk arising, either positively or negatively, that may affect the life cycle of a project. After decades of civil unrest, community water projects in the Republic of South Sudan have continued to face many challenges and risks associated with the management of the projects. This study aimed to determine the risks involved with water safety due to mismanagement in administrative duties. It also determined the risks the community faces due to the lack of governance and outdated approach.

To conduct this research, a mixed research method that involved both primary and secondary data sources was used. A purposive sampling method was used, utilizing semi-structured questionnaires and interviews for data collection to analyze risk management in the community water kiosks in South Sudan.

Thematic analysis of data produced shows many factors affecting the management of community water projects. Some of these risk factors are associated with the ongoing political situation of the country. However, the most striking factors are competition among water service providers, poor quality equipment, political leader interferences, and limited finances, insufficient use of existing policies, incomplete project implementation, and poor governance. A further study needs to be carried out on water.

The successful management of risks in the water sector in South Sudan shall require holistic approaches that should involve all the stakeholders, strengthening institutional capacity, and enactment of better laws and regulations, including contract laws. In addition, using modern business tools and techniques, including cost management, operational management, risk management, and project management, can greatly enhance the performance of water kiosk projects. Further, water service providers could consider the Lean Six Sigma approach as one way to remove waste and flaws in the country's water operations.

Keywords: *Community water project, governance, management, managers, risk factors, risk management, policies, pumps, South Sudan, urban water corporation, vendors, water kiosks, water organization, water projects, water utilities.*

Introduction

This chapter presents an overview of the research and its key objectives. The study examined the risk factors involved in managing water kiosks in rural areas of South Sudan. As other young African countries, South Sudan has a population of 8.2 million, as per the fifth Sudan population and housing census in 2008. The census shows that more people live in rural than urban areas. Around 55 percent have access to improved water sources (The Republic of South Sudan; The National Bureau of Statistics 2020). According to the UN report (UN-GLAAS 2012), the land surface is covered by an estimated 41 percent of water supply, and the country receives an average rainfall between 500 and 2,000 millimeters per annum.

The study focused on the three different levels of project management: the project managers, kiosk owners, and water consumers or users. The study embraced both qualitative and quantitative techniques to analyze risk management in water projects in South Sudan. Studying risk management in water projects helps with possible remedies to avert major future problems in the management of water kiosks in South Sudan.

Research Background

Humans are entitled to clean and safe water both for use and consumption. In a broader sense, water is essential for the economic, cultural, and social sustainability of any community. Even so, the threats to freshwater providers and the different needs for the resource have affected various regions in the world. South Sudan is among many countries in sub-Saharan Africa with a huge shortage of clean and consumable water, therefore needing communal water kiosks to ensure safe drinking water. The rural area inhabitants depend on such water projects to satisfy their needs, so communal water projects should be managed adequately.

Different reports have revealed that some community water projects collapse after only a few years of operation in the community. The failure of such projects impacts both the community members and project implementers with a mission to complete. Even though community water projects have been implemented in these communities, most African rural areas have experienced water shortages. Due to poor management, these projects have collapsed (Oyebande 2001). Limited access to clean, affordable, and safe water has posed various health issues to community members in the affected areas. Residents in these areas have been exposed to waterborne diseases because they consume water unfit for drinking. Hence, community water projects should have efficient managers to avert failure. Water provision schemes such as water kiosks are needed in rural areas in most African countries for the sake of clean water access and enhancing standards of life (Oyebande 2001). Water stalls are vital resources in any municipality and can satisfy various needs. Such a project meets residents' basic needs, such as adequate access to clean drinking water. It also allows the project proprietor to make a living and create employment opportunities. One aim of proper water kiosk management includes dropping the struggle for this product to all public members (Dagdeviren and Robertson 2011).

Effective governance of public water kiosks can guarantee water safety and efficiency, substantially reduce environmental destruction, and prevent disease epidemics of various sorts. The study analyzes various risks involved in supervising public water projects in developing countries, with a specific focus on South Sudan. The risks are often natural, which renders them difficult to prevent, but they could also result from human factors, which are outcomes of human movements and actions (Effah Ameyaw and Chan 2013). Different public water projects struggle with the administration of facilities because of the persistent risks involved. These jeopardies are comprised of sabotage and poor communal sustenance (Oyebande 2001). The study reflects on how supervisors can strengthen the administration of community schemes by defining and analyzing the dangers involved in community projects and then evaluating and analyzing the risks systematically and finding better ways to solve such challenges.

Water kiosk projects are a significant part of the rural community that supports their living and sustainability. These utilities provide safe and clean drinking water to individuals lacking adequate access to the commodity (Bey et al. 2014). They are most common in regions where the government is yet to invest in piped water for every household. However, the majorities of water kiosks in rural African regions are nonfunctional and collapse after they start operating. Reports have been published of water kiosks collapsing several months after being launched in communities, collapses often attributed to problems in management and sustainability (Jimu 2008).

The research was conducted to examine and investigate the reasons behind the collapse of community water projects in rural Africa. This study focuses on the risks of project management and how they affect the proper functioning of the utilities. The focus of this research project is to provide literature that informs how community water projects were managed and the risks that affected their effectiveness. Previous research on water kiosks focused on a broader topic of project sustainability without touching on the dangers of managing these projects (Nzengya 2015). The study focused on investigating the risks of managing these projects to close the gap created by earlier studies.

A significant risk in managing community water projects has been their failure to involve community members in their operations. As a result, these projects do not consider community needs regarding these water kiosks, such as the location of utilities, the time of opening, and how kiosks are operated. Other factors, such as prices for the commodity, have also been a challenge in managing the projects as high rates affect program beneficiaries. Because of all this, some community water kiosks built to provide the community with easy access to affordable water to rural residents have not been able to do so.

The findings from this research project will be used in informing policy and decision-making, especially for the government, nongovernmental organizations (NGOs), and donors looking to invest in utilities (Ong'wen 1996). For example, the government can use findings from this research project to mitigate the risks and implement a plan that will serve and benefit people. The study aimed to determine the different risks involved in the management of water projects in rural communities. It has investigated some of the reasons behind the failure of some community water kiosks and how vendors can work to change this pattern of risks to provide accessible and affordable water to the communities.

Problem Statement

The problem this study intends to solve is the constant failure of water kiosk projects. The study was done by analyzing the risks involved in managing water kiosks in rural areas in South Sudan. It investigated the reasons behind the failure of many community water kiosks and how vendors can control the risks. It focused on understanding the extent to which these factors influence the success of community water projects and how to mitigate these risks to enhance the success of water projects.

Sub-Saharan African countries have had the least access to drinking water relative to the rest of the world for a long time. In the sub-Saharan region and other emerging countries, the state government and other nongovernmental establishments (both local and worldwide) have spent an immense amount of money annually for access to clean and consumable water. Notwithstanding the recent interventions to deliver water to the world, access to quality drinking water was an unbelievable target owing to the growing human community and absence of viability of current developments (Salman 2011). Shortly after the implementers construct the project, most water systems do not achieve the expected objective of supplying the community with safe, clean water. The problems linked with such schemes that contribute to their eventual collapse should be reduced to make such water projects more viable and operative. The threats linked with water schemes might be coming from various sources, varying from methods used in financial management, policy, community engagement, and program oversight. Obtaining adequate information about the risk features, particularly those expected to have a major impact on the water project's success can have a positive effect on the water project's reliability. In this context, this research examined the risk factors influencing the quality of water-based public projects throughout South Sudan's rural communities.

Purpose of the Research

The research goal was to assess the risk features that touch on water kiosk success in South Sudan's rural communities.

Research Aim and Objectives

The study aimed to analyze the risk features that can affect the implementation of water kiosk projects in South Sudan. This aim will be achieved through the following objectives:

1. To examine how fiscal administration duties pose a risk to public water safety
2. To measure the dangers posed by the community water kiosk governance and approach
3. To inspect how the methods in project administration can expose a community venture to significant dangers

Research Questions

The research aimed to answer the subsequent questions from the study:

1. How do financial administration practices lead to different challenges to water kiosks in society?
2. How does the administration's framework pose a risk to water kiosks in society?
3. Could public involvement help minimize the risk associated with urban water projects?
4. How can a community water venture be exposed to more threats through project management activities?

Significance of the Research

The regional and international public, business organizations, nongovernmental establishments, and dogma-based groups are increasingly disclosing their substance as a productive replacement resource that contribute 24 percent of gross domestic product (GDP) to the country's development. A struggling economy and a large populace in South Sudan have driven 65 percent of people to dismal poverty. Studying factors that affect the success and challenges involved with local public projects is vital to the projects, as it will help developers to understand how project schedules, scale, and implementation of investment assist in project outcomes, increasing public competitiveness and productivity. If a water plan leads to insufficient output and advanced risks, the administration should return to the early preparation stages to revise the concepts of the plan and set a realistic potential target.

Consequently, it is easier to notice the issues the project faces before it has started than to tackle issues once it is too late. In these situations, there can be a delay in the project timeline, which is essentially a waste of time and effort. Thus, authorities in the managerial units of South Sudan and other governments and employing agencies might use the study findings to reinforce their public relations initiatives to increase productivity and deal with the challenges of the plan, making them more productive. The federal government could similarly apply the findings in conjunction with other analyses to determine how to communicate with the community when carrying out outreach programs and campaigns. The results could be used to implement community-based infrastructure projects in partnership with other stakeholders to ensure effective policies. Researchers could also use this analysis to identify the gaps useful for future analyses.

The study is also essential when formulating environmental policies and rural planning perspectives. In this respect, it demonstrates the role that communities play in managing the important resources at their disposal, such as community water kiosks, among other catchment areas. Community participation is likely to reduce the risks that such projects are exposed to, and the installation of such programs alone is not the end and does not guarantee success; instead, their sustainability does. In a complicated rural setup where people still hold on to their beliefs and are change-averse, more communal participation is required for both structural and nonstructural interventions. Moreover, the study is essential from an economic and development point. The fact is that one of the biggest challenges the developing countries face is how to finance and sustain the Sustainable Development Goals (SDG) for water provision. A study by Rutten et al. (2014) revealed that the expense incurred by various states would be approximately $22.6 billion in a year, about 3.5 percent of Africa's GDP. The money needed for operations and maintenance alone would be around 1.1 percent of Africa's GDP. Therefore, the findings of this study would be essential in ensuring the value of community participation is upheld and the maintenance and operational costs of such projects are subsequently reduced. The research finding is also very useful in demonstrating the importance of community involvement in the process of maintaining water projects and ensuring that projects are sustainable. Besides, there is a strong correlation between access to consumable water and economic growth. Therefore, policymakers will also find the findings of this study critical when formulating economic blueprints for rural communities.

The Constraint of the Research

The investigation was aimed solely at the rural regions of South Sudan; other non-South Sudanese public water schemes have not been tested or assessed; thus, the research findings will not provide a detailed overview of all public water schemes in South Sudan. Foster et al. (2019) noted that case study analysis is not systematic due to a lack of statistical validity and vigor about data collection differences. In this case, for example, data could only be gathered in rural regions, and the findings in a simple analogy, such as an urban environment, would not be compared with other projects; ultimately, the prejudice is also noticeable in the nature and analysis of prior study analytical materials. In this situation, the researcher pursued credibility to prevent any bias that might affect the findings. Because the investigation used primary data, there were several situations where subjects might have been hesitant to disclose evidence on topics considered inappropriate, such as the expertise of administration in their educational and professional qualifications.

Nonetheless, the problem could be addressed by guaranteeing to the research participants that the data provided are intended exclusively for research goals by maintaining anonymity and privacy on the evidence they receive. Since the analyses will use interview questions, this form of information capture has the floor-ceiling effect (where respondents choose the highest and lowest values without considering the values in between the highest and lowest point), generating an outline of the questionnaire falling short of understanding the item at stake, giving incorrect responses. In specific instances, certain respondents will not return the survey forms. An interview time table will overcome these impacts. The results may be restricted by respondents' capacity to respond to questions accurately and identify different events correctly. It will not give the participants a standardized time to respond to the queries, and the investigator must offer the participants sufficient time to provide relevant evidence.

The researcher experienced deficiencies that arose from conducting data collection and analysis. These limitations may impact the findings and conclusions of this result based on how they are done (Allan 2012). The first limitation is the use of assumptions when collecting data from participants. The researcher will have to assume that participants have experienced risks when managing water kiosks in rural communities and have the correct information to answer the research question. Furthermore, the researcher will base this research project on the assumption that participants will provide accurate data without using any guesses on several interview questions (Allan 2012). In essence, the emphasis was placed on respondents answering questions they were sure about while ignoring those for which they lacked accurate information.

Another limitation of this research project was the issue of time constraints and getting the right participants for the study. A rural community is a generally busy area with everyone involved in the daily hustle and bustle. It means participants may not get the appropriate time to answer all the interview questions without interrupting their daily schedules (Allan 2012). As a result, some respondents chose to respond quickly to the interviews and return to their everyday life, thus affecting the accuracy of the collected data. Similarly, time constraints affected getting access to all participants, considering the study targeted kiosk operators, owners, and organizations. It was challenging to catch up with the owners and get their opinions on the research objectives.

Assumptions of the Research

The conditions mentioned above directed the investigation; the specified participants were a fair reflection of the entire population in all the categories studied, and those participants willingly provided details, deprived of the fear of data privacy violation. The analyses also concluded that all of the poll's inquiries were compiled and the respondents had sufficient time to review the data, offering the subjects enough time to reply to every query.

Literature Review

Introduction

Chapter two presents an analysis of theories and models very important to the research. The chapter also reviewed water policies in South Sudan, the regional water system and stress, water policies in the Nile basin region, and the traditional versus new water supply model. The chapter further discusses the evaluation of sustainable community-based water initiatives and concludes by evaluating factors that reduce the community water initiatives' sustainability.

Water Systems and the Origin of Water Problems in South Sudan

A handful of previous research addresses water systems in Eastern Africa, water supply policies in South Sudan, the effect of water distribution on economic and political environments, models for water supply in developing countries, and role of community and private water suppliers in ending water shortages.

South Sudan gained independence from Sudan in 2011 after several decades of civil war. Besides having adverse economic effects on the people of South Sudan, the conflict limited the capacity of institutions to provide essential services, such as water and sanitation (UNICEF 2019). The country has many renewable water sources. The White Nile originates from Lake Victoria in East Africa and flows through Uganda and South Sudan before merging with the Blue Nile in Sudan. The White Nile is the primary tributary of the Nile. The Nile, the world's longest river, has an average discharge of 2,830 cubic meters per second (Rutten et al. 2014). The river's basin covers Kenya, Ethiopia, Uganda, Rwanda, Burundi, South Sudan, Sudan, and DRC (Rutten et al. 2014). More than 27 million cubic meters of water flow through South Sudan annually, but South Sudan has not managed to harness the water to end its water shortage problems.

UNICEF (2019) estimates that only 40 percent of South Sudan has access to essential water and sanitation services. Such coverage is classified as "very low" by international standards. Rutten et al. (2014) indicated that about 80 percent of the South Sudanese population does not have access to improved sanitation services, while two-thirds of the population does not have access to improved water sources. South Sudanese are forced to walk several kilometers to find clean water, according to UNICEF (2019). Though the government has set up semi-autonomous institutions to increase water supply coverage, South Sudan predominantly depends on international aid organizations to meet its water needs, especially in rural areas (UNICEF 2019). Why is South Sudan unable to exploit its natural water resources to provide its citizens with enough clean water? According to Rutten et al. (2014), South Sudan is unable to utilize its water resources due to high population growth, high rate of evapotranspiration, inadequate or inconsistent rainfall, and uneven distribution of the water sources. UNICEF (2019) blames the under-exploitation of underground aquifers, long periods of political instability, and the fact that the Republic of South Sudan is barely a decade old, meaning its institutions do not have enough capacity for South Sudan's current water shortages.

In South Sudan, about half of the water the Nile discharges is used by natural processes such as evapotranspiration (Islam and Susskind 2015). It is the process by which moisture is transferred from land into the atmosphere. Evapotranspiration consists of two processes: evaporation of water from the soil and transpiration by plants. Most evapotranspiration happens in the expansive Sudd wetland (Rutten et al. 2014). The Sudd is one of the largest freshwater ecosystems in the world; the swamp covers approximately 57,000 square kilometers (Rutten et al. 2014). Moisture loss through evapotranspiration in the Sudd wetland is estimated at 11.4 billion cubic meters per annum (Islam and Susskind 2015). Rainfall accounts for less than 15 percent of South Sudan's natural water (Islam and Susskind 2015). The country experiences inconsistent rainfall patterns; most parts receive thirty to forty inches of rainfall per year (Islam and Susskind 2015). According to UNICEF (2019), South Sudan has not invested considerably in water storage initiatives, so it does not put most of its precipitation into meaningful use.

Besides the high rate of water loss through evapotranspiration, the White Nile has also not provided South Sudan with adequate water because the river is transboundary (Islam and Susskind 2015). The river has sources in more than four countries. The rate of water withdrawal in Kenya, Tanzania, Rwanda, Congo, and Uganda affects the amount of water South Sudan receives (Rutten et al. 2014). Also, South Sudan's activities will affect the amount of water that flows into Sudan and Egypt. Countries in the Nile basin often have conflicts over the river use (Swain 2002). Considering erratic rainfall patterns and the unreliability of the White Nile, boreholes and wells are the primary water source for South Sudan's rural population (UNICEF 2019). Geologists assert that South Sudan has many aquifers that could provide enough fresh water (Rutten et al. 2014). However, the aquifers are yet to be utilized due to limited information on their production potential and distribution (Rutten et al. 2014).

The Second Sudanese Civil War lasted from 1983 to 2005. The end of the civil war in 2005 was the precursor to South Sudan's independence in 2011. During the many years of conflict, the North controlled most of the resources. Consequently, most parts of the South experienced little to no infrastructural growth (Rutten et al. 2014). A few months after gaining independence, the South Sudan conflict broke out, pitting government forces against rebels. The conflict destroyed the gains the country had made in improving water supply coverage (UNICEF 2019). Before the conflict, the government and international aid bodies devised a framework to end water shortages in the country.

The political instability disrupted the implementation of the framework. Key players such as UNICEF shifted their attention from increasing water supply to providing critical services (UNICEF

2019). South Sudan experienced drastic population growth following the end of the Second Sudanese Civil War. The population increase is attributed to the return of South Sudanese refugees who had run to the surrounding countries during the civil war (UNICEF 2019). The drastic population increase exerted pressure on the existing water infrastructure, causing water shortages (UNICEF 2019).

Water Policies in South Sudan

The provision of water services is under the Ministry of Water Resources and Irrigation (MWRI). The ministry develops water regulatory guidelines, improves drinking water to make it safe, enhances the capacity of water supply institutions at the central government and state level, and sets water tariffs for public water initiatives (African Development Bank 2012). The current policies were developed after the South Sudan Comprehensive Peace Agreement in 2005; international development organizations, including USAID and the World Bank, helped with the plans (African Development Bank 2012; USAID 2013). Water supply planning and management functions in urban areas are performed by the South Sudan Urban Water Corporation (SSUWC), which is under MWRI (African Development Bank 2012). The Ministry of Housing and Physical Planning must approve private and community-based water initiatives.

The South Sudan Water Policy was developed in 2007 and

adopted in 2009. It guides water, hygiene, and sanitation initiatives in the country (Fernando and Garvey 2013). The central principle of the Water Policy is the recognition of quality water and sanitation services as a human right for all people in South Sudan (African Development Bank 2012; Fernando and Garvey 2013). Also, the policy recognizes private water suppliers and community development organizations as critical partners in achieving the goal of improving basic water coverage (Fernando and Garvey 2013). The policy recommends separating regulatory and water supply functions to increase rural water supply coverage, such that the two functions are not placed under one body (Fernando and Garvey 2013). According to the Water Policy, semi-autonomous institutions should carry out urban water supply functions (Fernando and Garvey 2013).

The water, hygiene, and sanitation strategic framework for South Sudan is also a critical guiding document for water initiatives; the framework was developed in 2011 (African Development Bank 2012; Fernando and Garvey 2013). The strategic framework's main purpose is to draw out a plan for increasing access to improved water in South Sudan. It also aims to turn the principles of the Water Policy into actual water projects (Fernando and Garvey, 2013). At the state level, the Ministry of Physical Infrastructure and Public Utilities coordinates water and sanitation initiatives (African Development Bank 2012). States have dedicated departments that oversee the implementation of the water, hygiene, and strategic sanitation framework (Fernando and Garvey 2013). At the county level, the local authorities have water departments to regulate water and sanitation activities (African Development Bank 2012; Fernando and Garvey 2013).

The government's leading partners in water initiatives include the African Development Bank (AfDB), the World Bank, USAID, and UKAID. The USAID's water and sanitation initiatives are guided by its Global Water Development Strategy and Transition Plan for South Sudan (USAID 2013). The two frameworks identify five critical areas the government and its partners should focus on (USAID 2013). The critical areas include: (1) increasing the capacity of water supply institutions, (2) improving water and sanitation access, (3) building partnerships with private and community-based organizations, (4) incorporating sustainability in the design and implementation of community water initiatives, and (5) imparting behavior change regarding water and sanitation in the local communities (USAID 2013).

The AfDB's (2012) action plan identifies four areas that should be prioritized: (1) constructing new water supply facilities, (2) rehabilitating dysfunctional facilities, (3) enhancing the performance of urban water supply institutions, and (4) building the capacity of water supply institutions.

Keega (2017) cautions that the MWRI and South Sudanese water institutions lack the capacity needed to end water shortages in the country. Most efforts have gone toward building more water projects and buying more equipment, overlooking the human resource problems of the water institutions (Keega 2017). The lack of human resource capacity-building programs in public water institutions has created a big technical gap between the institutions and NGOs (Keega 2017). Though NGOs are supposed to work hand in hand with government organizations to improve basic water coverage, the technical knowledge gap often forces NGOs to carry out water initiatives alone (African Development Bank 2012). Most public water institutions work under general civil service guidelines, and they have no clear job descriptions (Keega 2017). Fernando and Garvey (2013) recognized that coordination and collaboration between public water institutions and community development organizations and private firms are central to achieving South Sudan's water coverage targets. Furthermore, the capacity of public water institutions at lower levels of government should be expanded to preserve the gains the country has made in recent years (Fernando and Garvey 2013).

The AfDB asserts that South Sudan's water problems can partly be blamed on inefficient institutional arrangements. The roles of some water institutions and directorates overlap due to the lack of clearly-defined functions (African Development Bank 2012; Matoso 2018). For example, the MWRI, SSUWC, and local governments all regulate urban water supply. The capacity of water institutions is also limited by unclear water assets ownership arrangements (African Development Bank 2012). The central government owns most of the water supply equipment, despite the recommendation of the Water Policy that semi-autonomous water bodies control such assets (African Development Bank 2012; Matoso 2018). Institutions such as SSUWC lack financial independence. SSUWC only retains 20 percent of what it collects from water consumers; the rest is remitted to the national treasury (African Development Bank 2012). The lack of financial independence limits SSUWC's ability to invest in big-budget water initiatives without seeking external support (African Development Bank 2012).

Oxfam cites inefficient and overlapping regulatory mechanisms in the water sector as one of the main barriers to private water investment in South Sudan.

The research demonstrated that the WASH sector in South Sudan remains embryonic, and while some guiding policies exist, legal gaps remain, and operationalization has been slow. This has led to a lack of clarity over sectorial roles and responsibilities, leading to fragmentation and duplication of such roles, poor accountability, and weak oversight systems—all of which can limit the sustainability of water service provision. (Matoso 2018, 8).

Some of the water and sanitation services assigned to the central government's institutions are also assigned to local and state institutions, which causes overlapping of regulatory functions (African Development Bank 2012; Matoso 2018). Furthermore, foreign private firms looking to invest in South Sudan's water sector must grapple with the lack of legal counsel on various compliance requirements (Matoso 2018).

Reasons for Implementing Water Kiosks as Community-Based Projects

Studies conducted in low-income urban centers and rural areas in Ghana revealed that community water services are crucial in filling the supply gap left by formal/state water providers. Siemens Stifrung (2015) insisted that community water services can generate a functional amenity that is flexible and can sufficiently deal with the challenge of water access; this is contrary to the more fixed, obstinate distribution technologies used by the national water supply systems. Wendl (2016) urged policymakers in Ghana to acknowledge community water provision as a sustainable option. As a result of devolution in the provision of water services in South Sudan, community systems are viable. Reducing the cost of community water project implementation due to the reduced fees of connecting grid electricity in South Sudan is an economic advantage for community water projects, especially in the planning and implementation phase. Since community water projects are small-scale in nature compared to the national water supply network, this will increase their probability of financial sustainability. Other advantages of community water projects are that they can be demand-driven and dynamic in accordance with local conditions. The vital success factors of community water projects have been known to include low operational costs, ease to supervise, and independence from subsequent utilities such as energy.

The Safe Water Enterprise (SWE) is an example of an innovative small-scale system that could be used for water treatment in different communities and has so far succeeded in the business. Siemens Stiftung, a charitable arm of Siemens, developed the program. SWE offers a model for the provision of drinking water in rural and low-income areas in emerging economies. SWE provides an end-pipe water treatment system to consumers in different communities. The only mandate for the individual communities in this system is to ensure the water is stored safely. The approach integrates both low-cost and low-maintenance water provision technology with an entrepreneurial approach. The model anticipates the collection of sufficient revenues that would cater to the operational costs of the project while simultaneously providing affordable water to the community and low-income households. In a bid to find solutions to the challenges that various community-based management units face, SWE projects are given to community-based organizations, while the daily operations and management of the water project are given to an entrepreneur who runs the utility on behalf of the community members or community-based management units. These entrepreneurs help the project attain sustainability, and at the same time, it helps them mitigate or prevent potential project risks. The SWE approach advocates moving away from voluntary community management units to more professional water service provision as a solution to risk management and sustainability. Figure 1 shows a summary of the SWE approach. So far, the project has proved to be successful in managing various risks associated with community projects and infrastructures, though the use of entrepreneurs has denied community members a sense of project ownership.

Figure 1. Community-Based Water Management Unit

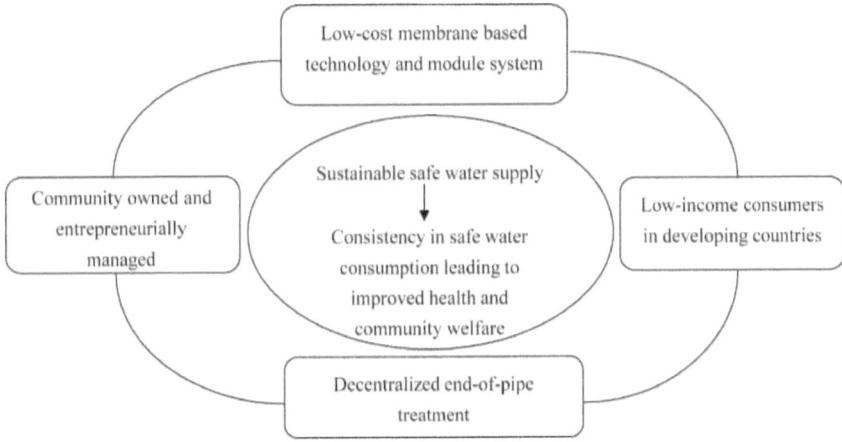

Source: S. Stifrung. "Safe Water Enterprises." Retrieved January 7, 2020. https://www.siemens-stiftung.org/en/safe_water_enterprises.

Various scarcity examination inquiries concluded that an effective strategy to be taken into account when formulating poverty alleviation techniques is to improve water amenities. Most communities have viewed a community water project as a one-time activity initiated only by private institutions. Such insight has resulted in poor preservation and has compromised the basic idea of first making the venture (Wendl 2016). While choosing the best development solution, community-based techniques have always been in dispute.

Current claims have run counter to this idea, stressing community support in enhancing their health and the further benefits of this methodology (Siemens Stifrung 2015). The outcomes of such studies have allowed community engagement as a developmental ideology to evolve into a fundamental issue when scheduling public development agendas. Richmond (2019) contended that community-based development (CBD) and its supplementary variant, the community-driven development (CDD), have developed one of the utmost favored approaches for channeling support to a group of people. In simple terms, CBD generally refers to the strategies that directly involve the recipients/beneficiaries in the process of project planning and supervision, while CDD, a term first used by the World Bank, leans toward the CBD project, which directly involves the community through management and regulation of various resources and supervision of investment capitals (Richmond 2019).

The general acceptance and implementation of CDD happened mostly because these methodologies were seen as approaches that could improve a community's sustainability, improve efficacy, reduce poverty, lead to community empowerment, and use a complementary factor to the marketplace and other public acts. The fundamental disagreement is that any communal scheme with an objective to attain the needed development is considered both a course and a result. Oino et al. (2015) stated that water ventures in developing states such as Kenya, Uganda, and South Sudan do not only emphasize attaining material resources; instead, they also aim at satisfying the desires of a community. Hence, such a project is intended to enhance the economic, political, and social schemes in the public. Dell'Angelo et al. (2016) argued that development projects identical to water stalls are distinctive in that they focus on democracy, consultative outcomes, equality, and integration. Whaley and Cleaver (2017) perceived that development schemes as a trajectory to community advancement could improve community welfare. The implementers should recognize the group's cultural, social, and environmental allegiances to avoid any unwanted risks ranging from lack of cooperation to destruction when carrying out development projects. Such aspects are critical in the affected regions for the comprehensive growth of society and civic mobilization.

Carter and Ross (2016) researched South Africa on the public change approach in 1994 and concluded that public projects are expected to enhance income-creating plans and the growth of small businesses in the public domain. Such projects are also likely to boost the rates of return of indigenous people, increase the accessibility of resources to all members of the public, generate job prospects, and advance the standard of living of individual community members. Oshionebo (2019) added that the utmost vital part of local schemes, like water kiosks, is that it makes the community autonomous rather than overly reliant on donors and international aid organizations. A study by Oshionebo (2019) reported that a project that meets the needs of the community highlights the division's financial, political, social, and cultural needs and makes the group autonomous when it comes to their resources, and the community is expected not to experience economic challenges as opposed to a community project that does not concentrate on such results. A social improvement would be viewed as the mechanism for this thesis in which disadvantaged people are encouraged to gain self-confidence and be inspired through projects. Oshionebo (2019) stated that community outreach programs imply rebuilding the current state of affairs and addressing the group's problems. The mechanism is complicated, multi-sectoral, and mono-faceted, with several critical industries to be considered. Other fields involve but are not restricted to poverty alleviation minimization and financial growth warrant.

The Essence of Water Stands in South Sudan

Water booths are designed to improve access to reliable water sources for people living in South Sudan. More importantly, accessibility to clean water is one of the Millennium Development Goals (MDG), currently rebranded as Sustainable Development Goals (SDG), and drawn by specific charitable organizations and nongovernmental entities to various claims. Silvestri et al. (2018) anticipated that sub-Saharan nations would not be able to satisfy the universally agreed water supply standard by 2040. It is necessary to incorporate the input of different investors, particularly those involved in the implementation and monitoring of water kiosks, for funders to accomplish the goal of obtaining safe drinking water. Emmett (2000) suggested that the active involvement of the intended beneficiaries in the preparation and administration of the water stands to be important when implementing such projects. Giving prospects for clean and consumable water in rural areas is a profound challenge, as it is not easy to determine a professionally structured policy to ensure that professional and fair supervisors enforce, maintain, and operate water kiosks. In rural areas of South Sudan, the population is not stable financially; therefore, the critical capital used to build water kiosks is usually given to the government or quasi-governmental organizations. The fact that the public cannot pay for these schemes and is not involved in any way has led to the situation that these water projects do not belong to the market; rather they belong to the government or institution. This lack of involvement of the public in the preparation and administration of the water projects has led to the perception that the project belongs to agencies in charge of project implementation. Such thoughts have elicited feelings of irresponsibility and unaccountability from the communities that host such projects, which has subsequently exposed these plans to more jeopardies. Akosa et al. (1995) stated that getting safe and consumable water in rural communities has helped both males and females conserve time and participate in other revenue-generating actions. In this way, consideration should be given to the need of both men and women in the design and implementation of community water supply projects to mitigate any risks that may be linked with the scheme. Harvey and Reed (2003) hold that most ventures of such capacity tend to ignore female participation.

Case Example of Previous Public Water Projects

Ghana was among the first African states to present a larger-scale community water source method, which is aligned with the current guidelines on water provision in different countries worldwide. The method used by Ghana involves the formation of local collections to the scheme, management, and execution of water schemes, which demonstrated to be more cost-effective than the preceding arrangement given by the ruling administration. The examination of public water missions in other countries like India, Benin, Nigeria, and Kenya exposed that applying community-focused methods immensely expands sustainability and decreases the jeopardies related to the water ventures. Kuper et al. (2009) analyzed rural water source programs funded by different institutions in forty-nine developing countries, indicating that community integration in these ventures is an absolute necessity for project potency and is expected to create public empowerment. When a population feels supported, they are less prone to deface a program, reducing the associated risks with these schemes. Davis et al. (2015) claimed that to advance transparency and minimize the risk related to collective water ventures, the groups involved ought to be involved in the operation and conservation of these ventures, acquiring the required replacement parts and making the appropriate changes to guarantee the project works correctly. Davis et al. (2015) recommended that the effective operation of public water schemes does not stop merely by deciding to cater for different expenses. Wendl (2016) observed the water and sanitation portfolio managers in Ghana and reported in most instances that community members did not recompense their yearly maintenance charges but also subjected these developments to even more dangers of depletion and lack of influence on the society as expected.

To prevent these patterns, Bonsor et al. (2015) suggested that society must have an agency or persons, such as treasurers, who will be entrusted with the responsibility to make rational decisions to ensure the community is committed and make various appropriate expenditures. At least a manager, director, and accountant should be present in any management team.

Nevertheless, in some situations and when the situation at hand requires it, certain roles could be developed to promote the public's effort and health. Water and hygiene leadership players in Ghana have been used as a case study to measure how community regulatory systems may reduce unnecessary risks. In Ghana, 30 percent of the resident authority is made up of females, and a community control committee is meant to keep the team responsible for monitoring (Hemson 2002). Bonsor et al. (2015) assert that some plans focusing on water supply in Zimbabwe's Zambezi Valley were not successful mainly because the groups around these schemes did not regard the unit as their project but regarded it as a foreign action plan; thus, they did not see it as their duty to protect the project.

The misunderstanding was significant even before creating

the project; the developers did not communicate with the local community. These circumstances left the community with an attitude they did not have any role in overseeing the project. The facilities were regarded as an open-access commodity because the initiatives completely lacked local ownership, which implied there could be no collective punishments or fines on those who damaged or defaced the waterways. There was also no restriction on the number of animals around the water source at the same time, there were no public management programs, and lastly, there were no security measures for public systems that are important in such projects.

The Zambian administration equipped several citizens and granted them government allowances in terms of wages. Such people were generally viewed as agents of government rather than members of the public. When the state stopped paying salaries due to the collapsing economy, the local inhabitants were no longer willing to contribute funds to support these individuals; nor did they have the required skills and experience to preserve water pumps since nobody cared to train the community members. Another unacknowledged factor was the pump operators' previous delineation. After multiple water efficiency and replacement tests, it was determined that adults are mandated to collect water in the community and should be educated on how pumps are to be operated. In a practical sense, the responsibility for collecting water was left to the children. Almost all of the faucets used were too heavy for one child, usually requiring the help of four to five children. The children were not informed of the water project's destination, although they traveled the longest distances from their household to the learning institute as well as from the houses of their families to the water catchment zone. According to Kleemeier (2000), the primary goal is to return people in the community to the water services' primary management and decision-making processes. The solution seeks to end the perception that only experts can give the best information on addressing a particular issue. Harvey and Reed (2006) argued that the public involvement course expands verdicts on the deployment of diverse water points to reduce possible dangers; the general population must be engaged in the sampling of the water project site, the technique to be used, the expenses to be contributed, and the administration strategy to be utilized across the entire procedure. Goldman (2007) argued that if the group is to be involved in the project, it will be essential to decide on the commitment plans in terms of costs and operations. If the public is to have a sense of belonging, this would decrease any uncertainties that might be correlated with the scheme. In certain situations, the community does not necessarily own the project, but being included in the project is essential to ensure the installation of required spare parts and implementation of any measures to protect the site from disruption and destruction of property. Community members ought to also be educated on how these thrusts should be maintained. Usually the state takes the initiative to repair the pumps, which were previously broken. Most of these improvements, however, were minimal. Administrations have taken an extended time to repair the project in some cases since some areas are inaccessible. Training water projects' primary group users, primarily women and children, can significantly reduce the potential risks.

Concerning technical differences and improvements, it is logical to conclude that different operating authorities need to decide on the sort of motors that both women and children can use with no additional support. The implementers might use community engagement to test the designated pumps in diverse sections to ensure the functionality of such schemes and develop the best fit for the community. Research results undertaken to study the impact of public support on the output of the Kiserian Dam in Kenya showed that the community had very little involvement in the process of designing, monitoring, and implementing the dam venture (Rutten et al. 2014). It even further inspired the efficiency of this water project and increased the risks involved.

Other Projects

Some other systems, such as water booths, would allow lawmakers to budget for upcoming projects and recognize the projects' adaptability. A venture's feasibility can be measured by utilizing various sustainability metrics that can be either empirical or quantitative. In certain schemes, the determinants used for valuation are not homogeneous, resulting in subjective auditing and assessment and lack of depth. Furthermore, integrating key factors of feasibility and risk analysis into project management inspection systems is necessary. Most systems are currently assessed and managed by using a plan intended to have a predefined theoretical sustainable development predictor such as practical, social, environmental, and social and legalities. A similar context was also applied to identify several water programs in Togo, Pakistan, and Malawi to realize the inherent perils and contemporary society involvement in risk moderation.

Lesson Learned from the Ghanaian Public Water Scheme

According to the assertions of the rural change advocates, community-based schemes are designed to achieve a particular function. For instance, this means starting and provoking growth urgencies from the marked public vesting and the political skills of the vulnerable population through including nonprofits and encouraging community members to interact in the interests of their well-being. Nonetheless, water access in rural Africa has been minimal and continuously restricted due to the constraints associated with existing community leadership approaches. Feasible water kiosk ventures want internal collaboration and external assistance, as stated in the report by Whittington et al. (2009).

Internal Community Support

There is a necessity for a structured public collaboration component that is accountable for handling the day-to-day functional objectives while overseeing the implemented projects for the community members to cooperate. Davis et al. (2015) suggested these systems can be made up of about eight to ten participants elected by the community, taking gender and tribal aspects into consideration. The unit's roles and obligations should be well-established in a joint resolution that is obligatory for both parties. The unit must have a scheduled activity calendar that also involves regular public meetings to reflect on the progress, limitations, and expenditures of the initiative. Komives et al. (2008) cautioned that no community project had been effectively conforming to these requirements for two consecutive years, as much as some systems are supportive and affirmative. However, the assumption that participants were chosen based on their ethnic and economic interests has resulted in minimal community partnership. These disputes have led to the decline of the partnership between the executive panel and the public and severely impacted the outcome of public projects. Salman (2014) surveyed leadership unit representatives in various local projects and stated that in most contexts, though the scheme has regularly presented its aims to the public, community members have shown discontent with the approach of the supervision representatives. Carr et al. (2012) also underlined the deficiencies of the community engagement units, which reported that external help is needed to examine the performance of community engagement systems and monitor seminars frequently held in public.

The Role of External Support

For several stakeholders, public involvement and monitoring is a critical method of determining rural community water resources, policy, and community participation expertise. Community administration is still confounding and ought to be distinct. Fielmua (2011) claimed that public involvement is an element of resilience in that community involvement is necessary to gain productivity, efficiency, and equity, whereas public management is not. Community engagement encompasses a course of empowerment review aimed at empowering the community as units of policymaking. In other words, community leadership requires community involvement, which involves a bottom-up strategy where public associates engage in a straight line in their growth. Leaders of the selected public take full charge of administrative processes and uphold various roles within the venture.

Von Korff et al. (2012) found many sustainability-related issues and risks caused by poor community management. Many initiatives employing the community-based method typically do not happen instantly after authorization but may involve one to three years to complete, so it is often challenging to demand a full review of these initiatives from scheduling to post-implementation. Different explanations have been provided for the catastrophe of group-based projects like water kiosks, and these explanations differ among communities and schemes. Nonetheless, recent studies have shown it is possible to reduce the likely risks related to diverse water amenities, such as the absence of continuous communication between the public and outside institutions, like ruling administration or benefactor/ executing organizations.

1. Insufficient civic training courses for project management, problem-solving, standard procedure, and facility sustainability
2. Lack of community development in the society to perform regular maintenance
3. Lack of consistent mechanisms for maintaining civic transparency and accountability, culminating in a loss of trust and admiration for the community at large and restricted community involvement

Once the initiating organizations adopted the approach of engagement, their conceptual inconsistency was illustrated mainly by a broad extension of introducing community groups such as water kiosks and encouraging efficient delivery while paying close attention to community planning and project readiness. This is important because the implementers are always trying to save money and time, thereby training staff on managing capital. Moreover, Effah Ameyaw and Chan (2013) indicated that if the growth and capacity-building cycle are extended, the group is less focused on the plan and may not even attend assemblies. Also, confusion and collaboration between community members and outside implementers are likely to result in adverse outcomes, especially where the implementer is the government. Rathgeber (1996) argued that most executing participants, such as administration and quasi-governmental establishments, believe that society can appropriately manage the system, which is not always the situation. This understanding is often accompanied by the reality that these institutions do not consider people's needs when carrying out such a venture or do not have an ideological basis for maintaining and running a program. Understanding the value of the community engagement plan after the project is initiated is essential. However, implementers should set up the process and tools to minimize potential perils at the ideal time and not hesitate until the programs are subjected to some threats and then concentrate on protective measures. The enforcement agencies have stepped aside in other situations, leaving the security and program provision to contractors whose only role is to inform when appropriate, not help the community. It should be noted that while their position in community engagement is still not observable, this issue does not include local business organizations.

The local institutions' external community support is essential for evaluation purposes. As much as the society-based approach has increased acceptance in the past few years, it is essential to highlight the simplicity and feasibility of the projects being implemented and to establish mitigation mechanisms and bestow systems to ensure the service distribution will check the minimum performance objectives. Institutional collaborations can accomplish different types of applying service structures. Collaborations are typically influenced by various facets, such as the complexity of the collaborations required and the personal and public relationships among participants to achieve a specific objective. Such casual and organized links were not widely available in the ventures described above, and other actors were not actively participating in the cycles of the venture. The disregard of other members and stakeholders reflected a distant relationship and provided the communities with an illusion, which, during the delivery and post-implementation phase, culminated in weak community cooperation.

Thus, to minimize any uncertainties involved with the program, it is necessary to include certain stakeholders at the beginning of any mission. Jackson and Gariba (2002) recommended that project management would focus primarily on the following facets to endorse a paradigm that can be used for risk analysis:

- Program tracking and evaluation framework
- Project capability development
- Relevant practical and monetary initiatives

Contrary to this, private sector involvement can also be a substitute approach, and as Harvey (2008) noted, the need for such a system highly depends on the condition at stake, and therefore more study and investigation are required before any framework is adopted.

Richmond (2019) further gives an ideal definition of what the term "water management" should be. He stated that water management should reflect a pattern of actions that renders the water project eligible for community use. Therefore, water management could be perceived within the prism of any methodology, which can be economical, administrative, or social, which will aim at attaining one or more of the following elements:

a. Managing the quantity or enhancing the quality of water needed to attain a specific objective
b. Decreasing the loss in quantity of water as it flows from its source
c. Changing the time of use from peak hours to off-peak phase to make water distribution more equitable
d. Improving the capacity of the project to continue serving the community even when the commodity is limited

In the same regard, Richmond (2019) compared the community water management process with the art of building a house. The author asserted that before building a house, there is a need for guiding policies, codes of conduct, workers with specific skills, and the homeowner. Just like the process of building a house needs guiding rules, managing community projects also needs a set of rules because it entails policies and plans for guidance, teamwork, skills, customers, and water users. In short, this is a multifaceted affair. As evident in this discussion, it is clear that water resources management and risk management are interconnected. It is impossible to pick one way and not the other. Generally, communal water risk management can be perceived in terms of the policy framework, technology, and community involvement. They may include other issues like water-taping technology, managing community expectations, balancing the project losses, and community emotions. If any of these factors are not adequately considered, there is a possibility of increased project risks.

Ultimately, this is a step where community involvement is an essential facet in water production. As Eguavoen (2008) stated, in the current world, managing community water projects needs more skills and perspectives than engineering, human management, or practicing law. Community involvement is necessary to ensure the community project is equitable and sustainable, particularly in developing countries such as South Sudan, where water is scarce. According to data collected by UNICEF (2006), the sum of individuals who had access to clean water improved by 23 percent in African countries between 1990 and 2004. During the same period, the population increased by 52 percent, which created water dependency.

Water Issues and the State of Affairs in South Sudan

Allan (2012) noted that in 2012, more than 12 million individuals in South Sudan had no access to safe drinkable water. Later, a substantial amount of public and donor capital has been focused on introducing community water stores in specific areas of the country's rural areas. Nevertheless, these programs had a lack of capability and reach. Given the lack of funding and limited operational performance, the chances for increasing the reciprocal positive relationships with the state level were small. Different community projects were carried out in South Sudan from 2005 to 2011 and supported by international donors and administration. Such initiatives are beleaguered at addressing the immediate needs for community building and recovery in war-torn countries and undeveloped regions.

However, the primary concern was the water system's feasibility. Harvey and Reed (2006) described sustainable water resources as an asset planned and implemented to strongly support the various goals of society both now and in the future while maintaining the dignity of the environment and society. This perspective provides a comprehensive view showing the several areas affecting the sustainability of a community-based water plan that encompasses laws and policies, organizational frameworks, social aspects, advanced technologies, and community development.

Risk valuation schemes are designed in this survey to use multiple multifaceted metrics to investigate and supervise several water-based plans in multiple rural points in South Sudan. The evaluation comprises revising supplementary knowledge, assessments of systemic literature, and opinions of experts. The danger ratings are assigned based on the sub-indicators observed. Research by Eguavoen on public projects (2008) using a comparable method showed that 40 percent of the public-based water stalls deployed were considered comparatively viable, although they were viewed as young ventures that only existed for about one to three years.

Due to inadequate post-implementation approaches from both state and donor-sponsored agencies, low community development, and financial assistance, the low profitability was mainly related to the problematic institutional and sales performance. Hence, there is a vital necessity to establish post-execution techniques and designs to promote water ventures, both theoretically and fiscally, to guarantee the viability and accomplishment of the venture's objectives.

Conceptualizing Risk and Risk Management

A risk is an unforeseen occurrence that, if it occurs, will lead to a positive or a negative effect on the project objectives (Guide 2001). Muriana and Vizzini (2017) defined risk as the possibility of a loss, injury, or other harmful outcome. However, in some instances, risks can also lead to positive outcomes or uncertain results. According to Qazi et al. (2016), a risk is a chance that a specific adverse event will happen within a specified period or an outcome of some challenge.

Risk management is mostly defined as a systematic process of identifying, analyzing, and responding to risk (JISC 2008). It is also perceived as an organized and holistic approach fashioned to organize, identify, and respond to various risk facets to attain the project objectives (Bey et al. 2014). In the context of a project, risk management involves the identification of influencing facets likely to negatively affect the cost, timeline, and objectives of a project and the metrics of effects of potential risk and implementation of mitigation measures to reduce potential effects of both the expected and unexpected risks (Muriana and Vizzini 2017).

On the other hand, Barrett et al. (2015) stated that risk management involves systematically assessing risk areas and carefully determining how each should be treated. It involves a management tool that identifies various sources of risks and uncertainty in determining their effect and establishing the relevant management response.

Processes of Risk Management

Risk Identification

The first step toward risk management usually is through informal means and undertaken in different ways according to the structures established by an organization and the team managing a project. Risk identification, usually the first step in risk management, involves past experiences and the analysis of similar projects that have been executed before. In this first stage of risk management, the integration of tools and techniques could be applied to identify the risks associated with any project. Various identification methods can be used in different types of projects. Hopkinson (2017) argues that risks and threats associated with a specific project could be tough to eliminate, and if the identification process is done properly, it becomes easier to manage risks and take the necessary actions. The essence of risk management is that stakeholders can prepare for potential problems likely to happen during the course of the project.

It not only facilitates the anticipation of the expected risks but also helps the project implementers prepare for unexpected risks.

Risk Assessment

Risk assessment, the second stage, involves the analysis of the collected data during the identification phase. Risk assessment is perceived to be a stage where the risks identified are shortlisted, starting from those with low impact to those with the highest impact. Two primary techniques are used for risk assessment, qualitative and quantitative assessment. The qualitative technique involves analyzing the identified risks formally. In most cases, a risk register is used in the whole process (Sadgrove 2016). Usually, a risk register consists of the following sections: description of the risk, classification, connection of the risks to other facets, expected impacts of the risks, likelihood of occurrence, suggested mitigation approach, and allocation of risks to stakeholders.

The quantitative assessment starts by classifying risks as high, critical, or unmanageable, as per the project implementer's assessment and capacity to mitigate the risks. The essence of this methodology is to establish a magnitude of contingency that should be used so that in case the risk occurs, there would be sufficient resources, time, and capital to manage the risk.

Risk Response Planning

The phase of risk response planning involves the necessary mitigation responses by adopting the necessary strategies in response to the positive and negative outcomes of risks identified. At this point, the project implementer allocates the roles, duties, and mandates of every shareholder/partner in case the risk occurs, such that in the actual occurrence of the risk, the blueprint is already laid out.

Monitoring and Controlling

If a risk occurs, there is a need to monitor the progression of the risk and its impacts. In other words, risk monitoring means assessing the indicators of the risk and managing its effects so it cannot go beyond the unexpected levels, which could lead to unforeseen impacts that could be devastating for the project.

Identifying Risk Factors

One way of mitigating risks or preventing such incidences is by having a clear picture of what exposes these projects to various risks. As Richmond (2019) identified, an organization will build a communal project, but what determines if the project will be viable is the community members themselves. Therefore, a major risk facing community projects is related to the community members themselves. The best way to prevent the foreseen risk is by first understanding the community members, understanding how they use the water project, and then formulating the necessary policies that would ensure risk management. AfDB (2012) argued that the best way to understand the community regarding risk management is by identifying who the frequent water users are, when they frequently use it, and for what purpose so that when establishing community water management units, such factors can be considered.

The German Development Cooperation (GIZ) established various community water projects in different parts of eastern Kenya and later assessed the projects. The assessment findings would help in understanding the community when it comes to managing community water projects and their subsequent risk management.

General Characteristics of Samples

The GIZ assessment obtained their samples at the water kiosks, and the information was essential in pointing out the household members frequently using the water kiosk for domestic use, as the respondents confirmed. Therefore, from this insight, it is already clear that most people using community water kiosks in South Sudan mainly utilize them for domestic purposes. Thus, when formulating water management policies, such factors should be considered. In other words, to facilitate a sense of communal belonging and reduce the risks associated, the water kiosk should satisfy the interest of the majority, which the GIZ survey noted as domestic use. In the same study, most of the interviewed respondents were between eighteen and forty years old, as shown in Figure 2, and the majority were females.

Figure 2. Ages of Respondents

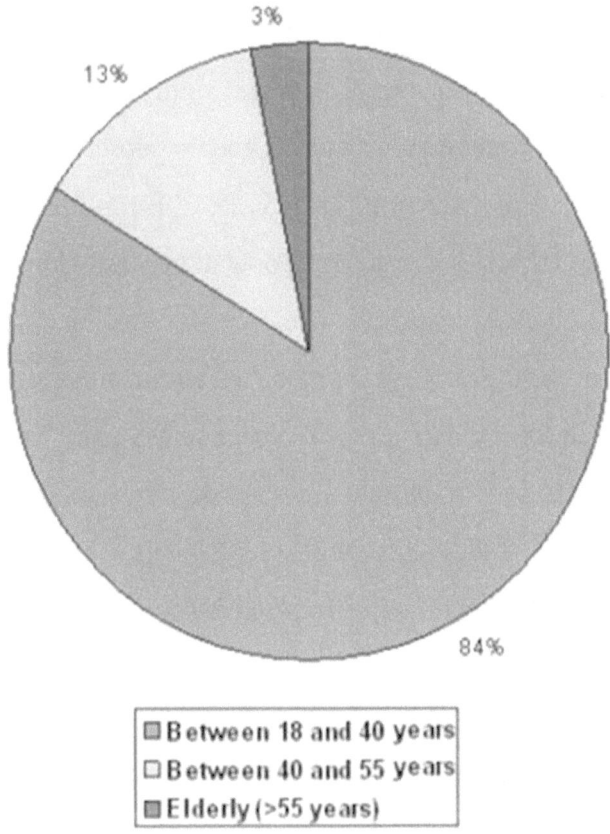

Source: EU-GiZ Report on Water and Sanitation

The study purposefully did not include children due to the long bureaucracy procedures associated with getting a permit that allows an organization to interview persons under the age of eighteen. Besides, research ethics instigate that a child should only be interviewed in the presence of their parents or guardian. In such a situation, it would be a challenge to interview children since most of the children collect water by themselves and are not accompanied by their parents or guardians. Even so, this study gives the necessary insights into the age group that frequently uses community water projects in sub-Saharan Africa. Both Kenya and South Sudan have a similar population structure, such that young people (people aged fifteen to thirty-four) are the majority, and therefore, a similar pattern should be expected in South Sudan (GIZ 2019). Additionally, most sub-Saharan countries have a large population of individuals between the ages of eighteen and forty. Therefore, it could be a coincidence that since this is the largest population age group in the area, they are also likely to form the majority of respondents in any project. If the same project were to be assessed in other regions such as Asia, South America, or the Middle East, the result is likely to be different, and the findings of GIZ cannot be perceived as conclusive and applicable in other regions or a different setup.

Figure 3. Population Structure of Respondents

(b) Gender of the respondents

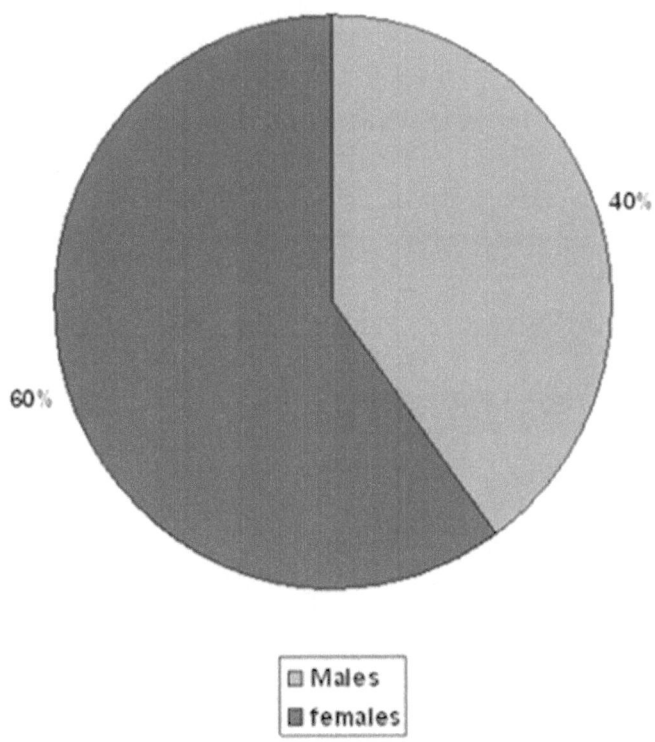

Source: EU-GiZ Report on Water and Sanitation

The information given in Figure 3 justifies the above argument that women need to be included in the community water management units since they constitute the majority of the community water project users. Secondly, according to the arguments of AfDB (2012), community water supplies are faced with huge risks during the peak hours of operation. That is the risk of not satisfying the needs of all community members and therefore forces the community water management units to establish various policies and frameworks on managing the resource during these periods to avoid exposing the project to further risks (GIZ 2019). The study by GIZ revealed the following periods as peak periods, as shown in Figure 4.

Figure 4. Peak Hours of Water Operations

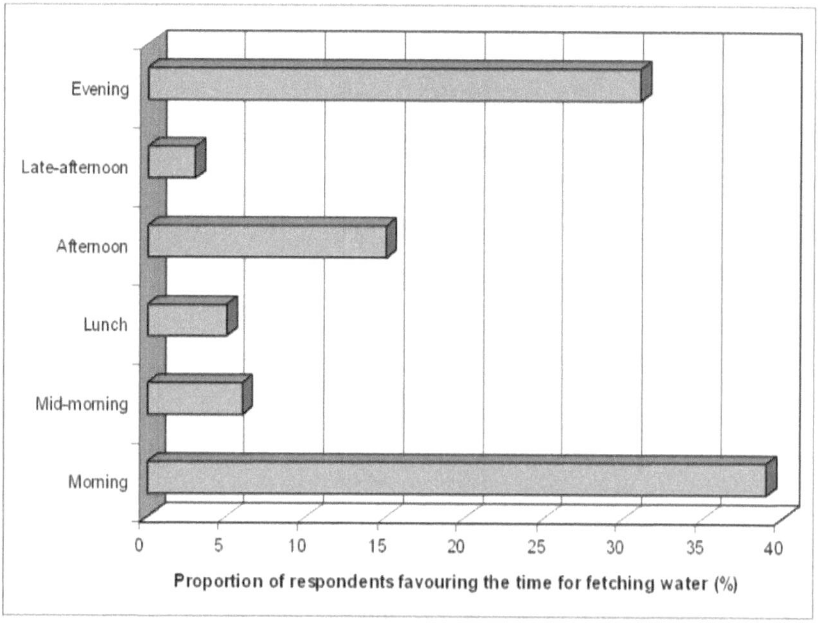

Source: EU-GiZ Report on Water and Sanitation

As per the information in Figure 4, the peak hours are clearly mainly morning and evening hours. The afternoon also had some significant attendance. However, the problem with the data is that it failed to give the specific hours and generally gave a range of morning, midmorning, lunch, and so forth. Such terms are not clear enough. Instead they should have considered giving exact timelines, such as 06:00 AM – 08:00 AM = morning. Nonetheless, the information provided is crucial for policy formulation and managing the whole process.

From the GIZ analysis, the evidence points toward the following conclusion in terms of understanding the community and establishing sustainable policies and guidelines that will reduce the risks associated with community projects. Community management units should give women a chance and establish a precise formation on how they will handle the community during peak hours (morning and evening) to avoid more risks.

Water Politics in the Nile Basin Region

Islam and Susskind (2015) and Rutten et al. (2014) agree that water politics, due to the transboundary nature of the Nile, are among the main reason why South Sudan is unable to exploit the White Nile fully. The average annual rainfall in Sudan is 10 inches, while Egypt receives between 0.0 and 7.87 inches of rain every year (Islam and Susskind 2015). The low precipitation forces the two countries to rely on the Nile. On average, Egypt uses more than 50 billion cubic meters of water from the Nile annually (Islam and Susskind 2015; Rutten et al. 2014). The amount of water that flows into Egypt fluctuates depending on how other countries use the Nile. It follows that South Sudan must consult other countries in Eastern Africa before constructing mega water projects along the Nile (Swain 2002).

Egypt struck several agreements with the Nile Basin countries during the colonial era. These agreements give Egypt the right to stop any project in other countries that it feels will reduce its water supply (Swain 2002).

A colonial-era pact between Egypt and other British protectorates in East Africa guaranteed Egypt more than 48 trillion liters of water from the Nile annually. The same agreement guarantees Sudan 4 trillion liters per year (Swain 2002). Another deal replaced the pact in 1959; the new deal involved Sudan and Egypt only. Under the 1959 agreement, Egypt's annual share increased to 55 billion liters (Swain 2002). Sudan's yearly share was adjusted from 4 billion cubic meters to 18.5 billion cubic meters per annum (Swain 2002). Most countries in East Africa oppose the two deals since they did not consider their needs. The upstream countries have started mega hydroelectricity and irrigation projects in disregard of the arrangements. Egypt threatens to go to war with any country that reduces its share of the Nile's water as guaranteed in the 1959 accord (Swain 2002).

Water Systems and Water Stress in the Nile Basin Region

Like in South Sudan, the primary water sources in Egypt and Sudan are the Nile and groundwater (Islam and Susskind, 2015). The southern neighbors, Kenya and Uganda, receive adequate rainfall per year and have many internal freshwater sources (Islam and Susskind 2015). Most parts of Kenya and Uganda receive more than

30 inches of precipitation per year. Like South Sudan, more than 50 percent of Sudan's population does not have access to improved water and sanitation services (Islam and Susskind 2015). Most parts of Sudan experience semi-desert conditions, and the country is yet to exploit the Nile to provide its residents with enough water (Makawy 2013). About 80 percent of households in Sudan depend on wells and boreholes for drinking water (Makawy 2013). Sudan has numerous wetlands that cover about 10 percent of its land. However, the wetlands cannot provide the country with enough water due to the high water loss through evapotranspiration (Islam and Susskind, 2015; Rutten et al. 2014).

The Nile and underground aquifers are the primary sources of water in Egypt. According to Rutten et al. (2014), Egypt draws roughly 56.8 trillion liters of water from the Nile annually. The Nile constitutes more than 90 percent of Egypt's total natural water sources (Islam and Susskind 2015). According to Sullivan (2002), Egypt is a water-scarce country. A country is classified as *water-scarce* if it does not have enough freshwater resources to meet its water usage demands (Sullivan 2002). Egypt's freshwater sources are hardly enough to meet its water demands (Islam and Susskind 2015). The country has four productive groundwater aquifers: the Nubian Sandstone Aquifer, the Mogra Aquifer, the Coastal Aquifer, and the Nile Aquifer (Islam and Susskind 2015).

Water stress scores are often used to describe the water shortage risks that countries face (Islam and Susskind 2015). The scores take into account the rate of water withdrawal and the available renewable water sources. Water is mainly withdrawn for agricultural, industrial, and domestic use. Water stress scores also consider the loss of water through natural means such as evapotranspiration (Islam and Susskind 2015). Ideally, the Nile Basin region should have low water stress scores (Islam and Susskind 2015). Sudan, Ethiopia, and Egypt have higher water stress scores than South Sudan does. Water shortages in Sudan and South Sudan are attributable to fluctuating water levels in the River Nile (Islam and Susskind 2015). The Nile's volume is mainly affected by evapotranspiration, runoff, and precipitation in upstream countries. Due to high temperatures, Sudan, South Sudan, and Egypt's net evaporative loss is negative (Islam and Susskind, 2015). A considerable amount of the rainfall received in the region is transferred to the atmosphere through evaporation and transpiration; a small volume runs off into water bodies and underground aquifers (Islam and Susskind 2015). Of all the precipitation received in Sudan, South Sudan, and Egypt, only about 10 percent can be used (Islam and Susskind 2015).

Tatlock (2016) indicates that water stress levels in the Nile Basin region are higher than in many areas of the world. He cites insufficient infrastructure as the primary reason for the high water stress levels. For instance, as of 2006, Africa had a total of 980 mega water dams, 590 of which were in South Africa (Tatlock 2016). In the same year, Tanzania had two large dams, despite having almost the same population as South Africa. Within Africa, the sub-Saharan

Africa region has the highest water stress levels (Tatlock 2016). Safe water coverage in sub-Saharan African countries is between 22 and 34 percent (Tatlock 2016). UNEP estimates the level of water stress and scarcity in sub-Saharan Africa will increase by 2025. Tatlock (2016) affirms that the solution to water problems in sub-Saharan Africa lies in having more water treaties and improving water supply infrastructure.

Unless countries in the Nile Basin region devise strategies to fully exploit their renewable water sources and harvest more water during the rainy seasons, their water stress risks will increase with time (Conway et al. 2005). Climate change, population growth, and economic development are among the main factors contributing to the increase in water stress risk (Conway et al. 2005). Rutten et al. (2014) projected that the human population in the region would increase from 400 million in 2015 to 700 million by 2040. The increase will stretch the capacity of the water infrastructure and renewable water sources. The Nile Basin region is made up of developing countries; their economies are likely to expand in the coming years (Rutten et al. 2014). Economic development is often characterized by urbanization, commercial agriculture, and industrialization. These factors increase the rate of water withdrawal (Rutten et al. 2014). Climate change and global warming will reduce the water available for use by increasing the rate of evapotranspiration (Conway et al. 2005). The region is also likely to experience changes in precipitation patterns due to climate change. This will lead to a decrease in the volume of water in surface water bodies and aquifers (Conway et al. 2005).

Traditional vs. New Models for Water Supply

The Inefficiency of Public Water Initiatives and the Need for Water Tariffs

Government institutions are inefficient in the provision of water and sanitation services; governments should therefore give more attention to facilitating a suitable environment for community development bodies and private firms (Fernando and Garvey 2013). In South Sudan, the mechanisms for private water supply are inefficient due to unclear standards and guidelines and untargeted subsidies (Fernando and Garvey 2013). Most South Sudanese civil contractors cannot construct high-budget initiatives due to the lack of financial and technical capacity (African Development Bank 2012). Foreign firms build most water projects in South Sudan (African Development Bank 2012; Fernando and Garvey, 2013). The cost of water projects is often high due to the lack of competitive bidding; this explains why South Sudan incurs the highest civil construction costs in Eastern Africa (African Development Bank 2012). Other reasons why South Sudan has high civil construction costs compared to other countries in the region are insecurity, inaccessibility of some areas, vast distances between the projects, and high cost of maintenance (African Development Bank 2012; Fernando and Garvey 2013).

Further, due to the lack of technical capacity, most local contractors build low-quality projects. According to Fernando and Garvey (2013), the cost of water projects in South Sudan can be driven down by introducing targeted subsidies, having a suitable environment for private firms, and leveraging public-private partnerships.

Gasson (2017) roots for enough public investment to drive down the cost of water projects, "Unless there is enough investment in public water initiatives, expensive water will become the norm" (African Development Bank report 2012, 2). Decentralization and privatization should be at the center of water policies, according to Goldman (2007), "Prices paid by water consumers in developing countries must rise substantially to avoid life-threatening shortages and environmental damage" (9). Budds and McGranahan (2003) maintain that government water supply institutions cannot achieve high efficiency due to the absence of profit motivation. "Central to keeping the pump working is financial viability. Achieving sustainable cost recovery sounds necessary and innocuous, but it has profound implications for those voiceless rural populations for which decentralization has removed the central state support, and poverty has denied private sector customer rights" (African Development Bank report 2012, 14).

In a World Bank-sponsored debate involving academics and water policymakers, the bank's proposal to push for more water tariffs in developing countries was supported unanimously (Goldman 2007).

Decentralization in Water Supply Systems and Community-Based Water Management (CBM)

In USAID's Global Water Development Policy and Transition Plan for South Sudan, local communities and different levels of government are the main partners in water initiatives (USAID 2013). Since local communities benefit most from water initiatives, they have more motivation to manage and rehabilitate the projects than federal and state bodies. The capacity of the community to start and run water initiatives should be enhanced by the other players (USAID 2013). USAID (2013) proposes several strategies for building the capacity of communities; they include assisting the communities in starting water initiatives, training, and having more significant roles for women in water management. In South Sudan, fetching water is mainly left to women and children (UNICEF 2013). Having women-run water initiatives can lead to more efficiency. USAID (2013) maintains that women have more knowledge of water location, improving water quality, and water storage.

Supply-driven and top-down water supply frameworks are unlikely to end water shortages in developing countries (Etongo et al. 2018). African governments, therefore, need to shun traditional water supply models and adopt innovative models. The inaugural UN's Water Decade was held in 1982; the need to introduce CBMs was the top agenda (Dodgson and Gann 2010). CBM models were also the top agenda at the Earth Summit meeting in 1992 (Dodgson and Gann 2010). CBM models aim to improve the community's role in managing rural water initiatives. They are premised on the proposition that the immediate beneficiaries of water initiatives should play the most significant role in their management (Etongo et al. 2018). After their introduction in the 1980s, CBM models have grown to become key features of water management frameworks in developing countries (Etongo et al. 2018).

Under community-based water management models, water management committees (WMCs) run water initiatives. The WMCs could be made up of up to ten members; the initiatives' beneficiaries vote in the members (Etongo et al. 2018). CBM models have not managed to end water problems in the areas where they have been implemented, despite being seen as the perfect replacement for traditional approaches (Etongo et al. 2018). About one-third of the water initiatives in sub-Saharan Africa do not achieve their objectives (Baumann et al. 2005). Etongo et al. (2018) place the rate of failure of water initiatives in rural Africa at 30 to 60 percent. In Tanzania, Etongo et al. (2018) reveal that roughly a quarter of community water initiatives do not go beyond two years of operation.

To improve essential water and sanitation coverage in Africa, Gasson (2017) puts forward a new system based on four fundamental principles: (1) incorporating local design in water and sanitation supply, (2) reducing the capital costs of water initiatives through decentralization, (3) spreading of risks among several partners, and (4) regarding quality water, sanitation, and hygiene services as a social guarantee (Gasson 2017). The undertaking of water initiatives should be in tandem with the goals of sustainable development (De Cecco 2012). Essential water and sanitation as a social contract entail viewing it as a human right that every citizen should have. When water is seen as a human right, key stakeholders focus their attention on increasing their coverage (Gasson 2017). Under the local design principle, Gasson (2017) insists that water resources belong to the local communities and the best way to manage such initiatives varies from community to community. Each community should develop its water management policies depending on situational factors (Gasson 2017). Franchising in water supply and water kiosks are tipped to end Africa's water problems by some experts. Gasson (2017) asserts they are not the panacea to water shortages in Africa, but they can provide an affordable solution in the short term. The Water Policy recognizes the critical role played by community-based management models in the water supply (Matoso 2018).

Governments and NGOs assist communities in running water initiatives; De Cecco (2012) is skeptical about the help. He argues that the government and aid bodies should have an exit strategy, such that the support is done within a sustainable framework. If the NGOs exit too early, the project may not be completed due to inadequate resources and poor leadership (De Cecco 2012). If the NGOs do not exit soon enough, the community will not have a sense of ownership of the projects, and they will not put in the effort to ensure the project's sustainability (De Cecco 2012). On his part, Therkildsen (1988) argues that the role of governments and NGOs in the provision of water services is unsustainable and inherently prone to inefficiency. According to Budds and McGranahan (2003), the functions of community development organizations in water and sanitation services should be limited to availing the resources necessary to keep the project running, enhancing public participation, and assisting the community to recover the cost of investment. Ideally, aid organizations should assist local communities with the financial aspects of water initiatives but leave critical sustainability decisions to the community (De Cecco 2012).

Most aid organizations in Africa use the daily availability of a certain number of liters of clean water, typically twenty, within one kilometer for every person to gauge the success of rural water initiatives (De Cecco 2012). Others assess the progress of water initiatives by the number of beneficiaries. The World Bank reveals there have been no considerable gains in improving water access in African countries (African Development Bank 2012; De Cecco 2012). Despite substantial investment by foreign aid bodies, most African countries have not halved the population that does not have adequate water and sanitation access in rural areas (De Cecco 2012; World Bank Database). As of 2012, the annual investment toward water initiatives in Africa by international aid bodies totaled $3.4 billion. The failure of the investments to considerably improve water access is attributable to the targeting of undeserving areas (De Cecco 2012; Etongo et al. 2018).

Evaluating the Sustainability of Community-Based Water Initiatives

Etongo et al. (2018) assert that ensuring the sustainability of community initiatives carries the same weight as starting them. Edwards (1997) states that sustainability measures how long the project will run after external leadership and financial support are withdrawn. Before embarking on a community water project, the sponsors should assess its sustainability. Several systems of predicting sustainability have been proposed in the past. Edwards (1997) proposes a sustainability evaluation model that takes into account four factors: (1) the availability of resources to cover repairs and the cost of operation, (2) the level of customer acceptance, (3) the physical soundness of the construction, and (4) the availability of adequate water all year round. The model developed by Carter et al. (1999) is also based on four sustainability indicators: (1) continuing support from the project's sponsors, (2, 3) recovery of the initial investment and cost of maintenance, and (4) the degree of support in the host community. The sustainability model that the UNDP and the World Bank use to evaluate the sustainability of rural water initiatives has five key indicators, as shown below.

Table 1. Sustainability Evaluation Model

Maintenance and Operational Practices	This evaluates whether the local community will be able to rehabilitate the project and keep it in good condition. It depends on the technical knowledge of the WMC and the accessibility of repair and maintenance tools.
Physical Strength	This measures the physical soundness of the structures. Factors considered under the indicator include pressure level, the quality of materials used, the quality of the structure, the possibility of contamination of the well, and whether there are leaks and defects.
The Willingness of the Community	This considers whether the local community supports the project.
Consumer Satisfaction	This considers whether the immediate beneficiaries are satisfied with different features of the project, such as the water quality and color, the hours and frequency of supply, the quantity of water available, and the taste.
Financial Management	This is concerned with the financial aspects of the projects. Factors considered are whether the community can afford the set tariffs, the criteria used in determining the water prices, and the financial management capability of the WMC.

Source: Mimrose et al. (2011)

Life Cycle Analysis (LCA), the sustainability analysis tool proposed by McConville and Mihelcic (2007), evaluates the sustainability of a water project at five stages: (1) identifying and analyzing the project's needs, (2) developing initial designs and carrying out feasibility studies, (3) final design, (4) developing an implementation schedule, and (5) maintenance and repair (McConville and Mihelcic 2007). LCA is illustrated in Figure 5 (flow chart).

Figure 5. Life Cycle Analysis

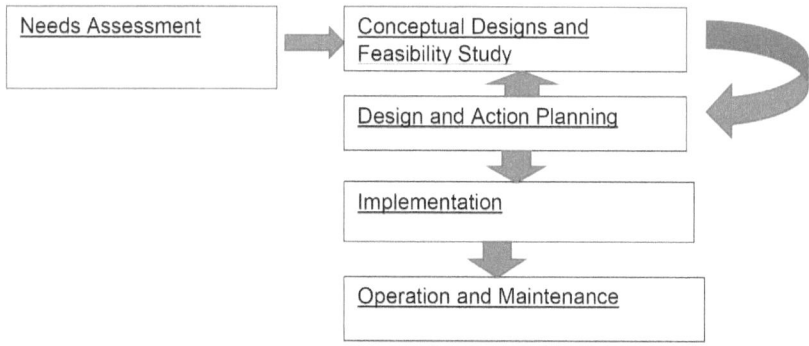

Source: McConville and Mihelcic (2007)

The double arrows between conceptual designs and feasibility studies and design and action planning indicate that the two stages can be iterated. In each of the above stages, five sustainability indicators are considered, as shown in Table 2 (McConville and Mihelcic 2007).

Table 2. Five Sustainability Indicators of Community Water Initiative

Community Participation	A major decision regarding the project's sustainability should come from the immediate beneficiaries. Community participation gives the community a sense of ownership of the project.
Economic Sustainability	The local community should have enough funds and technical knowledge to run the project without external leadership and support.
Political Compatibility	The water project should not contradict the host country's goals and priorities.
Social-Cultural Consideration	The water project should not interfere with the local community's culture and traditions.
Environmental Preservation	Nonrenewable resources in the locality should be used sustainably.

Source: McConville & Mihelcic (2007)

The most important sustainability pillars in LCA are economic, environmental, and social (McConville and Mihelcic 2007). While most organizations use the same factors to measure political and environmental sustainability, the criteria for measuring social sustainability vary from organization to organization. The UN considers solidarity, tolerance, equality, freedom, and shared responsibility when evaluating social sustainability (McConville and Mihelcic 2007). The Canadian International Development Agency

(CIDA) considers coherence, good governance, engagement of the civil society, donor coordination, and local ownership (McConville and Mihelcic 2007).

There are no standard tools for measuring the sustainability of community water initiatives; project sponsors apply multiple sustainability evaluation systems. Further, De Cecco (2012) asserts that sustainability assessment should be an iterative process. Generally, the sustainability scores reduce as the project ages. The scores are likely to be high during the initial stages and lowest in the later stages of the project (Etongo et al. 2018). Carter et al. (1999) maintain that a sound and quality construction does not guarantee sustainable water supply; rather, the structure and sustainability of a water project are different things. Carter et al. (1999) continue that it is more important to achieve sustainability and stability of a community water project in the initial stages. If sustainability issues arise in the latter stages of the project, the community and the project sponsors should have the necessary political, technical, financial, and social intervention mechanisms.

Increasing the Sustainability of Water Initiatives

Sustainability of rural water initiatives is the main goal in the South Sudan Water Policy (Matoso 2018). According to the policy, the central factor influencing the sustainability of the water initiatives is the degree of community engagement in the design, running, maintenance, and rehabilitation of the projects (Matoso 2018). On their part, Bey et al. (2014) list the physical functionality of the structures and capacity of the WMCs as the primary factors affecting the sustainability of community-based water initiatives. Bey et al. (2014) divide community-based water initiatives into two stages: the planning and implementation phase and the post-construction phase. The technical, social, political, and financial factors affecting the project's sustainability should be considered in the design and implementation stage (Bey et al. 2014; Matoso 2018). According to Matamula (2008), the most critical activities in the first phase of a water project are community engagement and training of the WMC. Bonsor et al. (2015) caution that community water initiatives will fail in the post-implementation phase if the WMCs are not trained on how to operate and repair the project. Therefore, project sponsors should conduct rigorous technical training before withdrawing (Bonsor et al. 2015). The training should focus on repairs, rehabilitation, and preventing water contamination.

Empowering the local community and ensuring the water project is used equitably keeps the project operational for many years (Etongo et al. 2018). Profits should not motivate community water initiatives, but the management should collect water tariffs from the community to run and rehabilitate the projects (Etongo et al. 2018). The management should explain the need for the tariffs to the community and criteria used in setting the tariffs. Matamula (2008) recommends that the community should select WMCs. The committee chosen will most likely be devoid of the requisite technical, leadership, and financial management skills (Bey et al. 2014). The project sponsors should train the WMC periodically (Bonsor et al. 2015). McConville and Mihelcic (2007) are skeptical about post-implementation external input; the community should control all aspects of the project after completion.

Aid organizations should never presume that WMCs and local institutions can make water initiatives sustainable; instead, they should assume they do not have the necessary capacity (Alaerts and Kaspersma 2009). When it is considered that the community cannot run the project by default, mechanisms for ensuring sustainability will be emphasized (Alaerts and Kaspersma 2009). Capacity-building should be done at different levels: individual, social, organizational, and civil (Etongo et al. 2018). Aid organizations should partner with the government to provide a suitable environment for private investors and aid organizations (Etongo et al. 2018). Education is one of the critical strategies for building the community's capacity.

Capacity development can also take the form of education, organizational improvement, and awareness creation and understanding of water use and value by local communities. Capacity development (CD) has been contextualized differently, and our understanding of the term has evolved from narrow conceptualizations of capacity, with a focus on individuals or organizations. (Etongo et al. 2018, 4)

Alaerts and Kaspersma (2009) define capacity development in the context of community initiatives as improving the community's participation and increasing their ability to gather skills for future use and learn from past experiences. It is described as the community's ability to create or manage sustainable growth (Alaerts and Kaspersma 2009). Alaerts and Kaspersma (2009) list four principles for developing the capacity of the community to run sustainable water initiatives: (1) increasing community participation, (2) equipping WMCs with leadership and human resource management skills, (3) creating an enabling environment through legal and regulatory frameworks, and (4) giving water institutions more resources and powers (Alaerts and Kaspersma 2009). The Thirst Project (n.d.) has an eight-stage framework for maximizing the sustainability of rural water initiatives in Africa: (1) community needs identification and qualification, (2) sampling, (3) hydrology and viability surveys, (4) pump test, (5) evaluating the water quality, (6) community engagement, (7) selection of WMCs, and (8) completing sanitation education. The stages have different purposes and activities.

Table 3. Eight-Stage Framework for Maximizing Sustainability of Rural Water Initiatives

Needs Identification and Qualification	In this step, the project managers learn about the characteristics of the local community. Valuable details include culture, financial ability, water problems, and whether the community has had water and sanitation infections in the past.

Sampling Area Wells	Sampling identifies the ideal location to drill the well. To identify the ideal site, the project contractors evaluate other water sources in the area to see the challenges they face. This stage also involves checking whether the area has natural water contaminants.
Conducting Hydrology Surveys	Hydrology surveys evaluate whether the underlying aquifer has enough potential. They also establish the nature of the aquifers so they can be protected during the drilling.
Pump Tests	Pipe tests measure whether the proposed well can provide enough water all year round.
Water Quality Tests	Underground water may be clear but unsafe for human consumption due to natural contaminants. At this stage, water samples are taken from the proposed location and sent to water safety experts for tests.
Engaging the Local Community	Community participation instills a sense of ownership in the beneficiaries. The community can participate in the project by contributing funds or sweat equity. Sweat equity refers to the effort, such as helping the contractors to clear the construction site.

Forming Water Committees	Project sponsors facilitate the selection of water point management committees by the community. The duties of the committees are to collect water tariffs and ensure the project is in good condition. Before withdrawing from the project, the sponsors equip the committee members with technical, leadership, and business management skills.
Hygiene and Sanitation Education	Before handing over the water project to the water point management committees, the sponsors should organize seminars to educate the community on water preservation and sanitation habits.

Source: Thirst Project (n.d.)

UNICEF (n.d.) uses a five-step sustainability model in its water initiatives in Africa. The five steps are iterative: (1) understanding sustainability, (2) planning and creating sustainability partnerships, (3) acting for sustainability, (4) tracking sustainability, and (5) programming the organization's processes to enhance sustainability. UNICEF (n.d.) maintains that sustainability depends on context and situational factors. There is no standard description of sustainability. Consequently, project sponsors must first describe it in the context of the project they are implementing. Understanding sustainability also involves setting sustainability goals (UNICEF n.d.). Contextual sustainability can be determined by conducting baseline and sustainability assessment research (UNICEF n.d.).

Placing sustainability in the specific project's context helps the project implementers discover potential challenges and develop a sustainability plan (UNICEF n.d.). As De Cecco (2012) indicated, water initiatives will be sustainable if they are in line with the host country's goals and priorities. Partnering for sustainability entails finding out the host country's water goals and preferences and working with government agencies and NGOs toward common goals (De Cecco 2012). Sustainability audits should be regular (UNICEF n.d.).

UNICEF (n.d.) cautions it is difficult to make generalizations regarding the factors that influence the sustainability of community water initiatives.

Different factors can affect different scales or levels of the intervention. For example, the absence of a coherent national WASH policy is critical and can be an obstacle identified at the sector level, while at the local subnational level, it is the actual knowledge about capacity, and practice of implementation of a policy that matters. The complexity of the sector can often make generalizations difficult: some authors identify as many as 25 factors affecting the sustainability of rural water supply. (Thirst project n.d, 19)

The most critical sustainability factors at the community level are the quality of service provision and the soundness of the physical infrastructure (UNICEF n.d.; Haysom 2006). UNICEF does not support funds and sweat equity contributions by the community, as the Thirst Project (n.d.) proposed. UNICEF (n.d.) asserts that community participation in water management should be within demand-responsive frameworks. Nevertheless, meeting the user needs is not enough to guarantee sustainability; the project sponsors should consult the community when drawing out the performance standards and let the community participate in monitoring the sustainability of the initiatives (Etongo et al. 2018; UNICEF n.d.). Further, the water project should consider the water quality, the environmental impact, the durability of construction materials and tools used, and the credibility of procurement procedures. The project sponsors, in collaboration with water point management committees, must define the performance standards; for example, water should be of desirable quality, and the determination of water prices should be transparent (Etongo et al. 2018). The guiding principle in the design phase should be sustainability, and it should be evaluated throughout the course of the project (De Cecco 2012; Etongo et al. 2018).

Factors Reducing Sustainability of Community Water Initiatives

A considerable percentage of communal water initiatives fail, but many research conclusions give no general reasons as to why. The sustainability of community water initiatives is mostly depends on situational factors, so it is challenging to give generalizations (Etongo et al. 2018; UNICEF 2019). Nevertheless, some reasons for failure appear in multiple pieces of literature. Failure to repair water supply equipment seems to be the top cause of failure. It is listed as the cause of failure in the studies by De Cecco (2012), McConville and Mihelcic (2007), Moon (2006), Nkongo and Tanzania (2009), and Etongo et al. (2018).

Whittington et al. (2009) investigated the relationship between the willingness of community water initiatives' beneficiaries to pay for the service and stalling of water initiatives due to a lack of repairs. This multi-country research considered different community-based water initiatives in Africa and South America. Only initiatives that failed due to a lack of repairs were studied (Whittington et al. 2009). The initiatives were carefully selected to ensure there was no possibility of failure due to low-quality physical structures and equipment (Whittington et al. 2009). The failed initiatives had active water point management committees, and the initiatives' beneficiaries were paying water fees monthly. The failure of these initiatives challenges the conclusions of Etongo et al. (2018), Goldman (2007), and De Cecco (2012) that having active water point management committees and charging the community water tariffs would guarantee timely repairs.

Charging the community water tariffs should help community water initiatives recover the cost of operation and have enough funds for repairs. For instance, according to Haysom (2006), availability of funds has the highest degree of correlation with whether a broken-down project will be repaired or not. While Haysom (2006) focused on water initiatives in rural Tanzania, Whittington et al. (2009) showed that charging water tariffs does not guarantee financial sustainability. Despite different positions on whether charging water tariffs ensure the sustainability of community water initiatives, Mimrose et al. (2011), Haysom (2006), and Whitington et al. (2009) affirm that most initiatives fail because of inadequate resources to carry out repairs. Nkongo and Tanzania (2009) explain why water initiatives lack funds to do significant repairs despite charging water fees. She reveals that most of the money collected by water point management committees does not stay within the project due to a lack of regulatory mechanisms. Nkongo and Tanzania (2009) focused on water initiatives in East Africa. Mimrose et al. (2011) performed research to establish the top reasons why community water initiatives fail, and their study focused on Sri Lanka. The failure of WMCs to repair the projects emerged as the top reason. The initiatives' beneficiaries were unable or reluctant to submit their monthly water fees, thus resulting in the shortage of funds to carry out repairs.

The lack of quality community participation is also cited in several studies as a top cause of failure of community-based water initiatives (Mtinda 2006; Hayson 2006; Mimrose et al. 2011). Most initiatives engage the community through the contribution of funds, recovering the cost of operation, giving community members roles in water point management committees, and having the community contribute sweat equity. However, Mtinda (2006) and Haysom (2006) caution that such engagement is not enough; it is challenging to achieve genuine community participation. According to De Cecco (2012), the form of community participation that most initiatives lack is involving the community in making critical sustainability decisions; external players mostly make these decisions. Other causes of failure of community water initiatives are a shortage of hands-on skills to condition and repair water equipment, weak physical structures, inability and unwillingness of the beneficiaries to pay tariffs, and poor transport infrastructure (Etongo et al. 2018; Gumbo 2004; Mimrose et al. 2011).

Town Overviews

This subchapter presents findings from the case analysis of community water supply in South Sudan. This analysis is based on field observation, semi-structured interviews, and a review of South Sudan community water supply project documents. The study covered water kiosks in Tambura, Yambio, and Juba Counties in South Sudan. The presentation of the findings begins with an overview of the town, location and communities living in the area, community water kiosk projects, utilities, and payment mode. The last part of the chapter concludes with the challenges and identified risks that the community water management projects in the country face.

About Tambura County and the People

Tambura County is situated in the Western Equatoria State of the Republic of South Sudan. Western Equatoria is a state in the southwest of South Sudan. Central Equatoria State borders it to the east, Democratic Republic of Congo (DRC) to the south, and Central

African Republic (CAR) to the west. Western Bahr Al Gazal and Lakes State lie in the north.

As one of the oldest towns in Western Equatoria State, Tambura County has an estimated population of 56,000 inhabitants (South Sudan National Bureau of Statistics). However, the population has increased over time, and there hasn't been any census conducted after the 2009 housing and population census. According to the Fifth Sudan Census 2009 (https://ssnbs.org), Tambura County is a good example of the development challenges the country is facing in both urban and rural areas. Access to water and sanitation is about 40 percent in Western Equatoria State (www.ssccse.org). In Tambura County, it is even lower, and the county lacks a reliable and effective water supply system. This situation demonstrates why it was important to conduct the research and assess the risk involved with the existing community water supply.

According to the government hierarchy, a county is the third level of government. It is headed by a county commissioner, the head of security who is either appointed by the governing body or elected by the people. However, the constitution of South Sudan recommends an elected county commissioner leads the counties (South Sudan Interim Constitution 2009, Amended 2011). Tambura County is geographically located at 4 degrees north and 28 degrees east, at an altitude of 640 meters above sea level. The area receives between 1,200 and 2,200 millimeters of rainfall annually. The Azande communities predominately inhabit the county. Other communities who migrated and settled in the county due to employment and displacement include the Jur and Bari. The Azande ethnic group forms the largest population of the communities in the county. The area is predominately inhabited by farmers whose livelihood depends entirely on agriculture and small trade. The county is surrounded by many small seasonal rivers and streams on which the communities depend for water during the rainy season. The area enjoys an estimated eight months of rainfall and four months of dry season. During the dry season, communities face the problem of clean drinking water. They move far distances in search of clean drinking water. The existing water supply systems in the area do not satisfy the needs and demands of the population in the area.

Tambura County Water Supply Projects

Three water supplies exist in this town. All of these water supplies are within the central part of the county. They were built by development partners with support from donor agencies such as the United States Agency for International Development (USAID), World Vision International (WVI) Australia, and Catholic Agency for Overseas Development (CAFOD). This research targeted two water projects: Tambura West Water Supply and Mamenze Water Supply projects. However, attempts were made to collect secondary and unstructured data from other water projects to enrich the data analysis.

Tambura West Water Supply Project

This project was constructed thanks to the collaboration between the United Nations Office of project services (UNOPS) and Women and Youth Empowerment (WOYE) with funding from the United States Agency for International Development. The role of Women and Youth Empowerment was to mobilize and train the community, form water management associations, train water user groups, and document and report project experiences, achievements, and challenges. The project was established through a feasibility study conducted to ascertain the community's willingness to participate in the management of the water supply. A grassroots-driven approach was implemented for the project by developing a project proposal and then constructing and handing over the water system to the community association to manage.

The overall project objective was to enhance access to safe drinking water. This was to be realized by establishing a reliable community water management system. This body manages water supply and distribution and sets a platform for a sustainable system by establishing rules and standards obligatory for water users and managers. The overall management of the water involved a broad spectrum of stakeholders, each with different roles and responsibilities, as indicated in the figure below (Water Supply and Management System Manual).

Figure 6. Broader Institutional Arrangement of Water Supply and Management

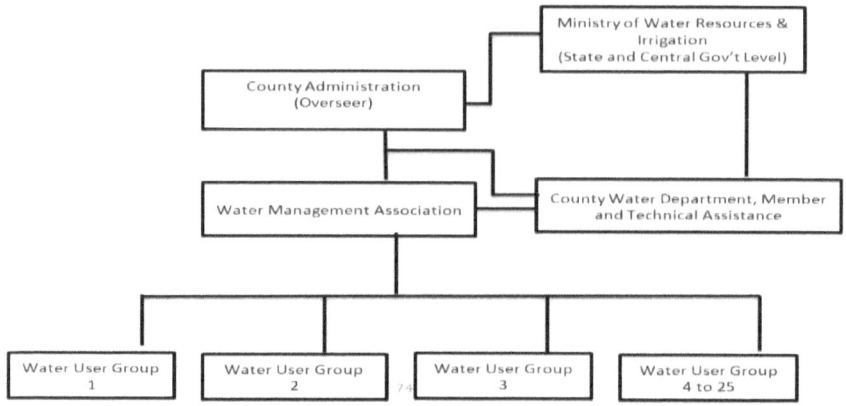

Source: Water Supply and Management System Manual (2012)

The water source, or borehole, supplies a minimum of 6,000 liters per hour, and the capacity of the tank is 258 cubic meters as per daily demand. The water supply is expected to last more than fifteen years before any major maintenance work could start. There is also a provision for future expansion to match the actual demand for water in Tambura County (TWWMP 2014).

Figure 7. Structure within Water Management Association (WMA)

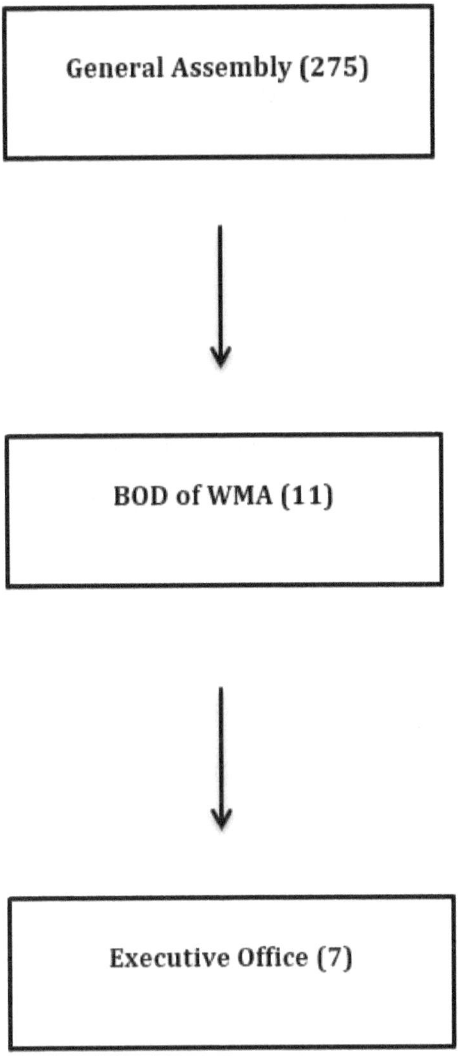

```
┌─────────────────────────────┐
│   General Assembly (275)    │
└─────────────────────────────┘
              │
              ▼
┌─────────────────────────────┐
│       BOD of WMA (11)       │
└─────────────────────────────┘
              │
              ▼
┌─────────────────────────────┐
│     Executive Office (7)    │
└─────────────────────────────┘
```

Source: Bonsor et al. (2015)

Duties of Water Management Association (WMA)

To discharge its duties, the WMA performs the following activities:

a. The Water Management Association shall be responsible for the management of the association and, for this purpose, may give directions to the office bearers regarding how, within the law, they shall perform their duties.
b. All money disbursed on behalf of the Association shall be authorized by the management of the Association.
c. The Water Management Association shall establish finance, procurement, and a monitoring/audit sub-association.

The Water Usage and Mode of Payment

To regulate the use of water, there are set rules and regulations. The Water User Group works in consultation with the local authority and community members. Members of the community are required to follow certain procedures in order to access the water. The set procedures guide the use of water to ensure there is better management of the facility. These procedures include:

- **Subscription:** Each household is requested to register with the Water User Group free of charge. Their details are entered into the record books, and cards are issued with details of payments each month.

- **Usage limit:** There is a set daily and monthly water usage limit. This also depends on the season of the year. During the dry season, water usage may increase, while during the rainy season, usage is low. To increase water use, the Water User Group and the Association, in consultation with the local authority, may arrange a meeting and make the decision.
- **Logbook of water usage:** The Water User Group keeps records of each household and maintains financial records. If the quota allocated to each household is met, the surplus would be charged.
- **Payments:** According to the project document, each household pays between 10 and 15 South Sudanese pounds. This amount was calculated in 2013–2014. This arrangement was reached during the consultation meeting, and community members suggested the amount (Bonsor et al. 2015).

Because of the economic crisis and the devaluation of the South Sudanese pound, the stated amount could no longer be used. The Association, together with the community, increased the fees to 200 SSP per household and 500 SSP for water point managers. This situation has been exacerbated further, and the fees no longer support the management of the water project.

Financial Management in the Water Sector

Water kiosk financial management is based on cost recovery and cost attainment principles. This principle is based on sound financial management of resource recovery and liquidity maintenance. Resource recovery aims to recover all financial needs, while liquidity maintenance emphasizes that all cash needs should be covered. The methods of achieving these concepts vary from country to country. However, in most developing countries, liquidity maintenance is important to realize permanent resource recovery and sustainability of the water kiosks (WHO report 2000).

South Sudan has made good progress in establishing an institution for water resources management in the country. The enacted and endorsed National Water Policy indicates government efforts toward improving the water institutions and financial management of the water sector. South Sudan government, the MWRI, and SSUWC, with support from donor organizations, work to ensure the water sector and sanitation are managed to provide better service delivery in the country. In order to ensure the sustainability of water, African countries have embraced the concept of water metering. This concept met several resistance from the water users, especially the poor communities who, for ages, have seen that water is free. However, this concept is now bearing fruits.

In this kind of arrangement, water users pay for the amount of water consumed on a regular basis. Payment is based on baseline findings conducted by the WMA, and user fees pass through various levels of water management. Treasurers collect the fees, keep records, and submit the funds to the Water User Group. The group does the verifications of the amount and pays the monthly charge to the operators. Also, the Water User Group deposits the amount into the account of the Water Management Association, as Figure 8 demonstrates.

The water sector financial management and liquidity should incorporate all the elements of production costs to avoid financial risks. The water sector should ensure there is cost recovery in the provision of services. It aims to maintain equilibrium in the financial management of the water. The average tariff should cover all the costs, that is, cost of goods and services, capital expenditures, salaries and wages of staffs, as this formula illustrates.

$$T_2 = \frac{C+M+P}{X} \text{ (market price)}$$

Where: C = capital expenditures

M = cost of goods and services

P = wages and salaries

X = volume sold

Figure 8. Sequential Flow of Financial Management

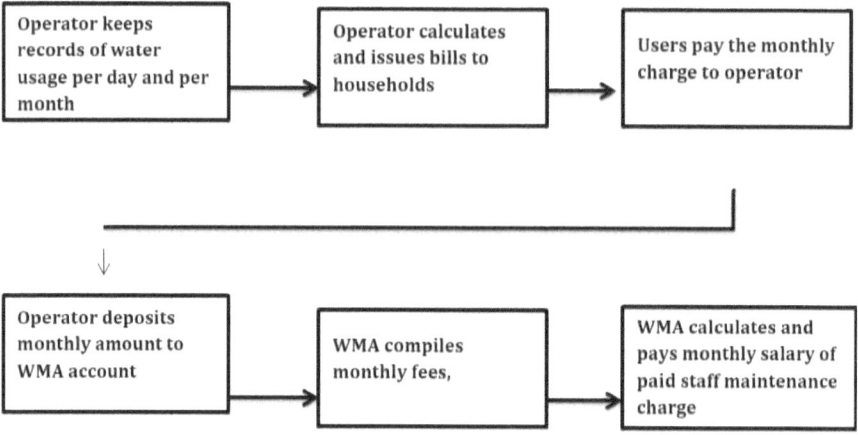

Mamenze Water Supply Project

Mamenze is commonly known as a water spring that has existed for many years. It is one of the water sources built in the early 1970s to supply water to the people of Tambura County. The project was established to supply water to a small population. But the town grew, and the water source became insufficient for the growing population. People traveled long distances to fetch water. After signing the peace agreement in 2005, many citizens of the county who fled to Central African Republic were repatriated back. The population became big, putting more pressure on the existing water facility. More residents started suffering from a lack of clean drinking water. In 2012, World Vision South Sudan sought to expand the facility to supply water to residential areas around the town. Two water tanks were established with the capacity to supply 9,375 residents of Tambura East Boma with clean drinking water.

About Tambura East Water Project

Mamenze water supply was established by World Vision South Sudan in collaboration with South Sudan Water Supply and Tambura County authorities. The project existed for only seven years before collapsing. There have been many challenges in leadership and resource mismanagement. Existing reports indicate there have been management challenges, which have led to the collapse of the facility. An unknown assailant killed the chairperson of the water group because he stood strong against the mismanagement of water resources such as solar panels and batteries. Because of this, many members became disappointed and left. The water supply collapsed, forcing residents to go back and fetch water from the springs. The established pipes were broken and blocked due to farm activities in areas where the pipes passed.

> According to the former commissioner, Hon. Charles Babiro Gbamisi, the unfortunate situation is that the water management has come to stand due to the breakdown of the pipes, basically the pumps and then the solar panels which were stolen during the crises in the county. So, these are the two things which

have made the water to be at standstill or breakdown. (INTRV #1)

However, the former commissioner is still optimistic that the water can still be functional if the solar panels and pumps are installed.

Establishment of the Project

In any project establishment, project implementation is important. The project officer follows the already-developed road map to establish the project. In the case of the Tambura East water project, the planning process and implementation did not go well. Two water pumps were proposed for the water project, but only one pump was installed, and the establishment of pipes did not follow the right path.

> From the start, it (project implementation) was not carried out fully as it was planned because we were supposed to have two tanks; one uphill behind the secondary school and the other one near the airstrip. It is only the one near the airstrip which was fixed. The one behind the secondary school was not implemented. This was due to the person who came to carry out the project program. That is why at the end of the project, we could not give him the certificate because he did not accomplish what was supposed to be done. (INTRV #1)

About Yambio County and Its People

Yambio is situated 440 kilometers west of Juba, the capital city of South Sudan. It is situated at an elevation of 650 meters above sea level. Yambio is the capital city of Western Equatoria State, one of the ten states in South Sudan. According to the 2008 census, Yambio had an estimated population of 106,000 individuals residing in the town. However, with the return of many refugees in 2009, the population has increased over time. The county has five payams: Yambio, Gangura, Li Rangu, Bangasu, and Nadiangere. The ethnic group inhabiting the county includes the Azande, Moru, and Balanda.

Before the installation of community water projects, residents relied on hand pumps, spring/river waters, and boreholes. This research study was conducted within Yambio County in areas where community water project has been established. These areas included Kpirabe, Naagari, Kuzee, and Ikpiro, and data collected were primarily from three of the six water projects in the area.

About Yambio Community Water Supply

Yambio is the state capital of Western Equatoria and one of the oldest counties. It has an old water supply facility through wells. After the signing of the peace agreement in 2005, many NGOs came in to support water and sanitation programs in the county. UNICEF, which has been in the county for a long time, has supported various rural water projects, including the supply of water equipment and rehabilitations of water pumps across the state. The establishment of the water project in South Sudan is primarily supported by many NGOs in partnership with the national government. In Yambio Township, at least two organizations, namely GIZ and World Vision, actively support community water projects. The GIZ WASH project is implemented in partnership with Yambio Urban Water and Sanitation Company (YUWASCO). The Yambio water and sanitation projects are implemented by NGOs with support from multilateral donors like USAID, GIZ, and World Bank, among others.

Kpirabe Community Spring Water Project

Kpirabe, a residential area within Yambio County with many inhabitants, is situated within 40 33' 44.39" N, 28o 24' 28.99" East 659 M ASL east of Yambio County. The project targeted 5,928 people, of which 1,300 are males and 1,500 are females and a total of 1,678 girls and 1,450 boys of the host community. Initially, UNICEF first developed this project in 1970, just before the Addis Saba peace agreement. It was later rehabilitated in 2006, but the project became dilapidated and posed a high risk to the community. The project has a capacity yield of about 8 cubic meters per hour.

Figure 9. Kpirabe Water Source

Source: World Vision South Sudan Program (October 2020)

Yabongo-Napare Water Project

This water project is located north of Yambio. The water source falls within the administrative area of Napere and can be easily accessed via Napere. The water source is not far from Ikpiro, but because of the population in the area, it is considered a separate water source with clear and distinct management. Just like Ikpiro water source, it has several water springs that feed the main river. The area is characterized by sand and silk content that floods could easily affect.

Ikpiro Water Project

Ikpiro water project is located north of Yambio and is easily accessible by road. The areas have several streams, but the most prominent rivers include Yabongo and Napere. The topography is characterized by surging soil and is minimally affected by floods. According to the geological survey conducted, the area has a medium potential of groundwater. The survey also indicated the water quality in the areas was expected to be good in terms of its color, turbidity, odor, and taste, though iron levels are expected to be slightly above the permissible levels.

Figure 10. Ikpiro Water Source

Other water projects in Yambio County include Kuzee and Naagori water projects.

About Juba County and the People

Juba, the capital city of South Sudan, is located in Central Equatoria State and has an estimated population of over 421,292 inhabitants. Data from the World Population Review (2021) indicates the town has experienced an increase in population by 4.48 percent growth rate. The city is estimated to grow more, as the urban population migration is steadily increasing in many African countries. The Nile River passes through Juba. Many people depend on water fetched from the Nile and transport it to residential areas using water tanks driven by vehicles and bicycles. The study covers two community water kiosk projects of Gudele water project and Gumbo water treatment plant.

Juba County, like any other county in South Sudan, is headed by a county commissioner. Because the town is the capital city for the state and South Sudan, many communities have settlements in Juba County. In fact, all thirty-two communities are represented in Juba. Also, many foreign nationals live here. Residents are mixed. They also have different beliefs, religions, tribes, and occupations. As the capital city of South Sudan, cohesion is challenging with communities slowly changing from a socialistic setting and embracing a capitalistic setting.

Juba Community Water Supply

The provision of clean and affordable drinking water to the residents has been a big challenge for the governments for many years. Many residents of Juba County depend on Nile water for their daily use. Despite achieving independence in 2005, South Sudan has been having problems with water in almost all ten states. The Nile River passes through the capital Juba, but still, there is an acute water shortage for quite a long time. The existing piped water, which was constructed in the early 1970s, could not sustain the increased number of residents. In fact, these pipes were mostly installed in government offices and residential areas. Many in the community depend on water being carried by trucks and bicycles but sold at exorbitant prices. Many boreholes and water pumps have been dug in different parts of the country. Some of these water pumps and boreholes are not well-maintained, putting residents at high risk of waterborne disease.

Due to this situation, several organizations have come in to support water projects in South Sudan. Most of these organizations are concentrated in Juba. Donor agencies, such as USAID, AfDB, JICA, GIZ, and the World Bank, among others, fund water projects in different parts of South Sudan. A number of water projects exist in Juba; however, for the purpose of this study, two projects were randomly selected for data collection, Gudele water project and Gumbo water treatment plan, both situated within Juba County.

Gudele West Water Project

Juba is a typical example where a lot of development partners or agencies dealing with water and sanitation are situated. Because of these, there are some elements of competition among agencies in order to deliver services. The demand for water remains very high. Several boreholes have been dug, and many water tankers carry water every day in all the suburbs of Juba, yet the demand remains high. This has prompted many development agencies to set up water treatment plants in and around Juba.

Gudele West is one of the suburbs of Juba County. It is located in Northern Bari Payam along Juba Mundri Road. Gudele is divided into three parts: Gudele One, Gudele Two, and Gudele West. Gudele West is one of the largest residential areas in Juba County. The water plant in the area consists of 80 meters-deep borehole water. It has an elevation of a water tower with the capacity of 30,000 liters of water, providing an estimated 5,000 residents with clean drinking water.

Gumbo Water Treatment Plant

Gumbo is Boma within Rejaf Payam. It is a vast area on the east bank of the River Nile in Juba County. It is characterized by a busy market hosting several nationalities from East Africa. This vast area is called Gumbo Sherikat. Many warehouses and companies are settling up there, making the area densely populated. The area is expanding, resulting in a high water demand. The average demand for water in Gumbo is estimated at 6,400 cubic meters per day. Gumbo is located in Central Equatoria State (CES) as a Boma in Rejef Payam. Juba has an altitude of 460 meters, with a maximum temperature of 38 degrees Celsius and a minimum of 20 degrees Celsius. The rainy season lasts between April and October. The water treatment plant provides water to an estimated 20,000 people. The treatment plant provides water to over fifteen water tanks for distribution around the county. There are also bicycle water vendors supplying water to small and medium restaurants and houses. The plant uses solar-powered water treatment to harness rainwater from the river, process it, and provide clean water to the communities.

Causes of Water Shortage in South Sudan

River Flow and Rainfall

The amount of water and evapotranspiration phenomenon is another reason why the 27 million-cubic-meter water of the White Nile flowing through South Sudan is not utilized as the ecosystem consumes a large quantity of that water flow. This river flows through other states like Congo, Tanzania, Kenya, Rwanda, and Uganda. These countries use the water, and what is left comes to South Sudan. After passing through South Sudan, the river continues into Sudan and Egypt, and they also have their specific water requirements. The dispute among the states over the water is a historical event in Africa.

Rain is scarce in South Sudan, and rainfall fulfils only 15 percent of the water requirement.

Return of Refugees

Another issue faced by South Sudan after the independence was that many refugees returned to their homeland. These refugees had taken shelter in surrounding countries, but after South Sudan became an independent country, they started moving back. This sudden increase in the population caused the water reserves to face a shortfall. More people means more water usage, and a newly formed country, which was hardly able to fulfill the needs of its current population, was not in the position to attend to the needs of returning refugees.

Local and International Bodies

Many local government bodies and international organizations are involved in the planning and implementation of the water and sanitation supply in South Sudan. On the local level, the MWRI is responsible for the water supply and irrigation at the state level. MWRI is responsible for making the guidelines, performing actions to improve drinking water, and installing new water resources. Two international organizations actively help the Ministry to develop policies: the AfDB and USAID. There are two documents of importance in this regard. The first is the South Sudan Water Policy, whose development started in 2007 and was implemented in 2009. The highlighted principle of this policy was that good quality water and sanitation were the basic needs of every citizen of South Sudan. This policy also welcomes private water suppliers and community developers to install new resources after consulting the Ministry of Housing and Planning. It was also mentioned in this policy that two different bodies should handle the urban and rural water supply management. In this regard, the SSUWC is responsible for water supply in urban areas and works under the MWRI. The second document is the Water, Hygiene and Sanitation Strategic Framework for South Sudan, formulated in 2011. There is a technical gap between the locals and the foreigners, which is causing a hindrance in the development process. There is an issue of human resource management between the NGOs and the locals, which must be sorted out to make the development process smooth. The local, private, and international bodies need to synchronize to achieve the desired results.

USAID's Findings

South Sudan falls under two USAID projects: Global Water Strategy and Transition Plan of South Sudan. The USAID team highlights five critical areas:

- The number of water supply institutions should be increased.
- Access to clean water and sanitation needs to be improved.
- The government should make partnerships with private and community developers.

- Sustainable design and implementation of community water initiatives are required.
- Awareness campaigns are required to change the behavior of the locals toward water and sanitation.

Findings of African Development Bank (AfDB)

The AfDB has also pointed out four areas regarding the water issues of South Sudan:

- New water supply sources should be constructed.
- The dysfunctional water supply facilities should be rehabilitated.
- The urban supply management process needs enhancement in performance.
- There is a need to build large-capacity water supply institutions.

AfDB also states that the inefficiency of the local and state-level water supply institutions has a great influence on the water shortage in South Sudan. As multiple institutions are handling the same areas, unclear job descriptions and definition of roles have caused a lot of overlapping, which reduces the performance efficiency by large. MWRI, SSUWC, and other local bodies are all managing the urban supply, which is causing problems. The central government controls most of the assets, and the bodies are not financially autonomous as prescribed in the Water Policy. The SSUWC gets only 20 percent of the revenue generated from water consumers, and the rest is submitted to the National Treasury, which restricts the Institute from the construction of new and expansion of old water resources without seeking assistance from other bodies. The AfDB has also pointed out the communication gap between foreign organizations and the local population.

The Significance of Community-Based Water Kiosk Projects

Following the footprints of Ghana's research regarding the community-based water services, it was observed that the national water supply leaves gaps that community-based projects can fill. After observing this gap, Ghana's policymakers promoted community water services, which resulted well for them. As the existing water supply management system is failing in South Sudan, the community-based approach is a sustainable choice. The national water supply is more obstinate and fixed, but that could be modified at the community level according to the requirements as community water services allow flexibility in the supply. As the country is developing its basic infrastructure, the economic efficiency of community projects also provides an edge. These installations can be easily connected to the national grid for electricity. The operational costs of these projects are also much lower than those of a national project, which is the type of financial sustainability required by South Sudan at the moment.

The example of SWE provides a lot of knowledge and direction. This enterprise works for the availability of pipe water supply in the rural areas of developing countries. They have only one criterion for the locals, that the water is stored in a risk-free manner. This approach ensures low costs and easy maintenance using an entrepreneurial method. This method is focused on generating revenue such that operational costs are at a minimum and the end user has to pay lower utility bills. This type of approach is required in South Sudan to solve the issue of accessibility and quality of water. Entrepreneurship is another approach that is the future, and even developed countries are obliged to use it for their sustainability. The opportunity for entrepreneurship is immense in a country like South Sudan. It will benefit the local residents and the foreigners who invest there by generating great revenues.

Risk Management

Risk assessment, prediction, and mitigation are significant parts of a successful project. As the water kiosk projects in South Sudan face numerous risks, it is important to introduce risk management tools to the system. An unforeseen occurrence that can lead to a positive or negative change toward a project is called a risk (Guide 2001). It is an unpredictable event that can have challenging or adverse effects during a project (Qazi et al. 2016). A loss, injury, or another undesired outcome can be categorized as a risk. However, sometimes it could lead to a positive change as well. Risk management can be understood as a process oriented toward identifying the possible events, which could cause a negative impact on the costs, duration, and aims of the project. It also includes planning mitigations and implementing them to reduce the impact of these risks (Muriana and Vizzini 2017). Risk management involves the creation of a process for the identification of possible risks and a response plan in case these risks occur in reality. Risk management is done systematically by predicting all the risks and designing a mitigation plan for each one of them separately.

Risk Identification

The first step of risk management is usually done through informal means and is undertaken in different ways according to the structures established by an organization and the team managing the project. Risk identification, typically the first step of risk management, involves an analysis of past experiences and similar projects executed before. During this first stage of risk management, the integration of tools and techniques could be applied to identify the risks associated with any project. Various identification methods can be used in different types of projects. Hopkinson (2017) argues that risks and threats associated with a certain project could be very difficult to eliminate, but if the identification process is done correctly, it becomes easier to manage the risks and take necessary actions. The essence of risk management is that stakeholders can be prepared for potential problems likely to happen during the project.

Not only does it facilitate the anticipation of the expected risks, but it also helps the project implementers prepare for unexpected risks.

Risk Assessment

The second stage involves the analysis of the collected data during the identification phase. Risk assessment is a stage when the risks identified are shortlisted, starting from the ones with the lowest impact to the ones with the highest impact. Two major techniques used for risk assessment are qualitative and quantitative assessment. The qualitative technique involves analyzing the identified risks in a formal manner. In most cases, a risk register is used in the whole process (Sadgrove 2016). Usually, a risk register consists of the following sections: description of the risk, classification, connection of the risk to other facets, expected impacts of the risks, likelihood of occurrence, suggested mitigation approach, and allocation of risks to stakeholders.

The quantitative assessment starts with classifying risks as high, critical, or unmanageable, as per the assessment of the project implementer and capacity to mitigate the risks. The essence of this methodology is to establish a magnitude of contingency that ought to be used so that, in case the risk occurs, there would be sufficient resources, time, and capital to manage the risk.

Risk Response Planning

This phase involves adopting the necessary mitigation responses by adopting the necessary strategies in response to the positive and negative outcomes of identified risks. At this point, the project implementer allocates the roles, duties, and mandates of every stakeholder/partner in case the risk occurs, such that in the actual occurrence of the risk, the blueprint is already laid out.

Monitoring and Controlling

If a risk occurs, its progression needs to be monitored, as well as its impacts. In other words, risk monitoring involves assessing the indicators of the risk and managing its effects so it cannot go beyond the unexpected levels, which could lead to unforeseen effects that could be devastating for the project.

Operations Management

After analyzing the data and examining the water situation in South Sudan, it was observed that operational management tools are not being applied to the full extent. Studies have highlighted that operation management tools are exceptionally effective in modernization and development. If the operational management tools are applied, the efficiency of the operations and the managers' performance are increased. As the operations are not managed in an optimal way, there are various backlogs, deficiencies, and wastes in the process of solving the water issue in South Sudan. Modern and proper operational management tools are required to be applied to remove all the waste and delays from the system.

Successful operations management requires minimum resources

to satisfy the end user. Operations management is an important part of project management responsible for managing every step of the process. In this case, the steps include creating new water kiosk projects, removing waste from the old ones, and delivering water to the end consumer. The operations managers are to handle the production, storage, quality, and supply of water. The project managers and operations managers should have the authority to make decisions regarding policies, design, quality, capacity, and funds/ assets. In this case, the SSUWC is responsible for the water supply to the urban citizens of South Sudan, but they only get 20 percent of the bill collection. A scientific approach to management is required in South Sudan. Scientific management has four major characteristics (Dell'Angelo et al. 2016). The first step is the development of a truth-based science. The second step involves scientifically-selected workers and managers. The third step is the training and education of the workers. This is especially required in South Sudan as foreign vendors and NGOs have trouble communicating with the locals. The last step is a compatible and synchronized work environment between the managers and the workers. When the local workers and managers can communicate well with foreign stakeholders, it will increase the efficiency of water projects and create a friendly work environment between the involved stakeholders.

Cost Management

Another important tool for the success of any project in this modern era is cost management. Once the waste and flaws are eliminated from the operations, the costs automatically decrease. Applying modern cost management techniques can further reduce the financial constraint of the new country. After examining cost management theories, it was easier to identify the areas where the costs of the water projects could be reduced. After going through various studies and cost management theories, it is safe to say that cost management can be categorized into four steps.

Resource Planning

In the planning phase, predictions are made based on expected events, and the estimations are prepared for the required resources. Economic, physical, and human resources are all estimated in this step. The estimations are made on the grounds of quantity, time, and other relevant parameters. Resource planning is initiated, and an execution plan is prepared. In this stage of project development, the work breakdown structure (WBS) and organizational breakdown structure (OBS) are developed. This is usually done for new projects, but it has been successfully applied to existing projects as well.

Cost Estimating

Costs and price estimations are made depending on the type of project. Technical knowledge is applied to finances with the help of different software-based techniques. The obtained results help greatly in making the business plan, analyzing the costs, and determining the total cost. As any project or cost can be divided into multiple cycles or stages, the cost of each cycle is determined independently. At this stage, more information is available, and estimations can be made much more accurately. The use of technology and software has allowed organizations to be efficient in their operations and management.

Cost Budgeting

After all the costs are estimated, it is time to allocate a budget for each part. The budgeting process allows for monitoring the project's progress and efficiency. A cost baseline is obtained by management after budgeting. A time frame is also determined for the cost budgeting schedule.

Cost Control

Cost baseline is selected as the reference point, and it determines what is going according to the allocated budget and which items are exceeding it. The variance between the estimations and the original costs are identified, and improvements are made to reduce the costs. In the cost control phase, two major things observed are performance and expense. The actual costs allow the management to determine the factors that introduce variance in the cost baseline.

Entrepreneurial Actions and Innovations

Entrepreneurial actions and innovations have proven to be consistently successful for project managers. The results of the projects implemented previously from different organizations have been examined and wisely included in this research during the process of planning and designing the action plan for water issue in South Sudan. Innovative strategies and policies are required to solve this water issue (Burns 2013). This will remove hindrances from the operations and increase the efficiency of the water supply management. According to modern theories of innovation, an innovation is a unique idea or imagination, which is converted into a workable plan for the purpose of gaining profits. Innovation can also be an update of an existing project to find a better way of solving a problem in accordance with market trends and actual costs (Dodgson and Gann 2010).

When something new is added to a project or a new kind of action is taken, it is safe to say that it is an innovative action. To solve the water issue in South Sudan, it is essential to use innovations. The model SWE offers is a good example of the sort of innovation required to solve the water issue in South Sudan. Innovations are helpful because they reduce the financial constraints of the new country and eliminate the operational and managerial limitations in water projects. In recent times, entrepreneurial innovations are proving to be more successful in generating more revenue and reducing costs

(Bessant and Tidd 2016). There is a difference between inventions and innovations; the latter performs the same actions but more economically and efficiently (Cohen et al. 2017). Entrepreneurial innovations are essential to solving the water supply issue in South Sudan.

Innovative methods should be included in the process of hiring and training the workers. There is a vital need to innovate urban planning and other logistical details to benefit from the available resources and reduce resource waste. The current structure of water organizations and suppliers was thoroughly reviewed for identifying areas where opportunities can be availed or created through an innovative approach. Entrepreneurial actions are also very important as they require minimum capital and are prone to be self-sustained. Creating new opportunities and identifying existing ones is also an entrepreneurial approach required in South Sudan to deal with its water issues (Rae 2007). Entrepreneurship has evolved over time, but according to the modern definition, it is a way of creating a business by utilizing minimum resources to generate profit. The rapid progress of technology has opened various new opportunities while making the existing ones easier. Millions of people use the entrepreneurial approach to achieve success with their projects and businesses. If we go further in entrepreneurship, it is acknowledged it is a plan of action for a specific project with economic objectives and deals with social, environmental, and health-related issues (Gibb 2000). There is a great need to train workers, managers, and vendors to make efficient communication possible between them. There is a big issue in South Sudan that the local and foreign stakeholders are unable to communicate properly. Entrepreneurial actions are required to set up training centers, especially for language learning for locals and foreigners. As South Sudan is facing financial constraints, it cannot allocate a huge capital for starting up new things required for the success of operations; entrepreneurial actions are the best choice. Instead of setting up a conventional building for a training institute, modern learning cards, apps, and verbal education are a better way. Setting up a virtual training center app would allow everyone to access the learning environment with minimum use of resources. If all stakeholders would be involved, this would mean thousands of individuals getting trained. If thousands of individuals use an app on a daily basis, there is a potential to earn profit through ads and affiliate marketing. In this way, the training process becomes self-sustained and might even generate profits.

Entrepreneurship is not just about generating profit, but also about finding solutions to greater problems. In South Sudan, the entrepreneurial approach is the best way to move forward in every field, including water management. One thing that makes entrepreneurs different from other people is that they see opportunities instead of limitations. An entrepreneur identifies existing opportunities and also actively thinks about creating new ones. This type of motivation and approach is the driving force behind the success of any project or action. It must be understood that there is no issue that cannot be solved (Piperopoulos 2012). In South Sudan, it is significant to implement an approach prone toward identifying existing opportunities and creating more entrepreneurial actions in the water supply and management of the country (Wickham 2006). Sometimes, the opportunities are already there, but they have not yet been identified. An entrepreneurial and innovative approach would enable the management to create new opportunities and remove waste from the process.

Summary of Literature Review and Gap

This chapter has reviewed various literature and underscored its effectiveness and influence on the research study related to risk management in community water kiosk projects in South Sudan. It began by analyzing water systems and water problems in South Sudan, detailing the policy issues, effects of civil unrest, and challenges associated with the River Nile in the East African region. This chapter also presented definitions of risks and risk management as a process that can result in either positive or negative effects. This process involves management tools in order to identify risks in community water projects. Samples of community water projects in sub-Saharan Africa were analyzed, looking at internal community support and the roles of external support to the projects. Reports from other studies suggest the management of community water kiosk projects is mostly done by females, as compared to their male counterparts. Other researchers who studied risks in water kiosk management focused more on sustainability and risk rating in water projects.

In the context of South Sudan, there has been no properly documented empirical work on risks and risk management in water projects. Given the country's distinctiveness and its civil unrest, it can be implied that management of water project risks can be a challenge. It has also been noticed that there are limited regulations and policy frameworks to guide the management of the water sector in the country. Further, insufficient water sources, especially during the dry season, pose some risks to the projects because of not satisfying the community members' needs. These different aspects of risk management are central to the study of how community water projects are managed, along with their effectiveness in the community. Thus, these aspects may contribute to effective risk management in water projects in South Sudan.

Research Methodology

Introduction

As shown in the following segment, this chapter presents the research design methodology and the data collection methods used to carry out this study. The method used explored or addressed the research questions and objectives of the study.

Research Design

According to Coldwell and Herbst (2004), research design is the glue that holds research together in perfect rigid shape, providing a framework to carry out the research in its entity. This study focused on analyzing risk management in community water kiosks in South Sudan. The researcher used the qualitative research method. According to Holloway and Wheeler (2002), qualitative research is like a form through which social inquiry is made, in which people share their thinking and experiences about the world. The study compiled various data and literature and analyzed them in relation to the research objectives. This approach was applied to analyze the existing links between policy, administration, fiscal, and public processes in determining risks facing different water projects in terms of objectives. A descriptive study strategy informs the researcher how to gather group data, which helps the researcher assess the respondent's present position compared to the parameters. It is also a suitable technique to extract positional data to define the research subjects, which would logically be too large to be explicitly studied.

Research Approach and Methodology

This research is conducted as an entrepreneurial action plan to obtain accurate results after cycles of planning and action. A research methodology is of great importance while conducting research. While developing the methodology for this study, the Saunders Research Onion was utilized (Saunders et al. 2016).

Figure 11. Saunders Research Onion

The visual illustration provided by the Saunders Research Onion allows the researcher to combine different approaches according to the requirement easily.

Research Design Option

The research design options chosen from different layers of the Saunders Onion are shown below.

Figure 12. Research Design Options

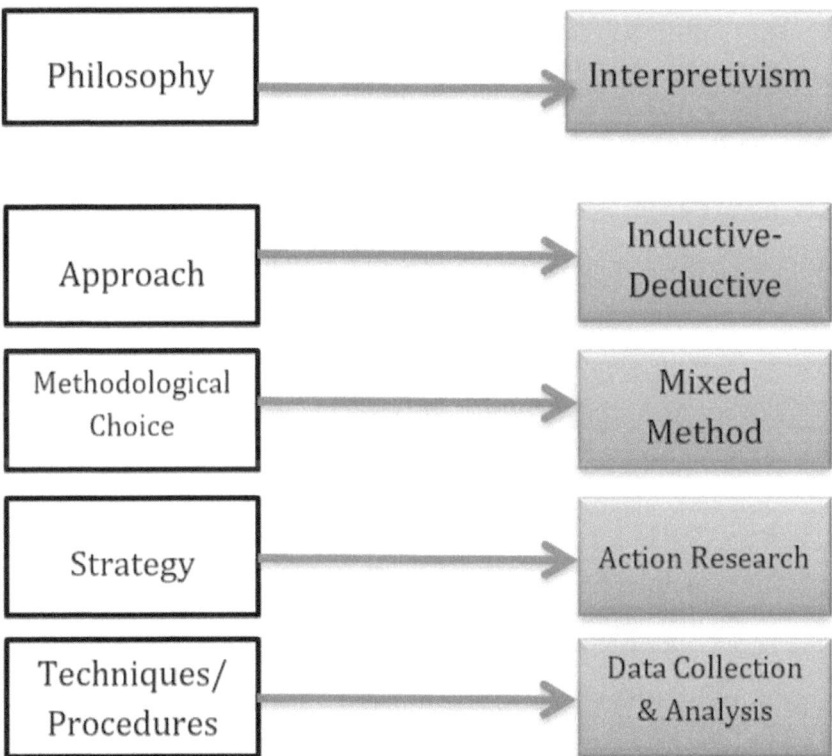

This study has been done by choosing a mixed-method descriptive correlational study. The qualitative and quantitative methods and techniques are used in a combined effect to determine the water issues in South Sudan concerning the local and foreign suppliers/ vendors. When both kinds of data are available to reflect the research questions, the mixed-method approach is the best choice (Morse 2016). This research used the quantitative approach to collect data related to water requirements, rainfall, revenue generation, and financial constraints. The design of the research is significant in every step of its progress (Walliman 2011). The qualitative approach was used to collect the data required to understand the historical background of the issue and social constraints. A mixed-method has allowed a deep understanding of the water issue in South Sudan and the development of an action-based strategy for best results. The analysis and presentation of the collected data are shown in the following sections. This study is correlational as it illuminates the relation between the water issue and local and foreign vendors.

Research Philosophy: Interpretivism

When the human interest becomes integral to research while focusing on different cultures from different points, such approach is known as interpretivism. The key stakeholders in this study are the consumers, water vendors, water suppliers, community builders, government bodies, and international organizations. The firsthand observation of the managers, vendors, and suppliers is also collected through three different sets of questionnaires to identify current procedures' flaws.

Research Approach

Traditionally, the studies were prone to adopting the deductive approach. With time, the inductive approach found its place among the researchers. After combining these two approaches, a new hybrid approach is formed, known as the inductive-deductive approach. This study has adopted this hybrid approach for the accuracy and authenticity of the research. During this research, when the identification of the issues and opportunities was of concern, the deductive approach was used. As this issue is a prevalent one and many entities have done experiments and tried different actions, inductive approach was also required to conclude the best findings and recommendations (Gibb 2000). It was also required to find out if all the facts have been covered in previous studies and actions or not. This approach greatly helped find the flaws and waste in the operations causing the water issue to remain unsolved until now. The results of the studies and actions taken by other organizations provided support to this research.

Implication of Lean Six Sigma Approach

In this approach, a collaborative team effort is required to improve performance and remove waste using a scientific method. Anything using resources but not contributing to progress or performance is considered to be a waste. Removing the waste from the operations is required to improve the water management and supply in South Sudan. The removal of waste not only results in the smoothness of operations but also reduces costs. The waste are categorized further into eight types (Gopalakrishnan 2010): defects, extra production, delays, unused human resource, transportation, inventory, motion, and over-processing.

Lean Six Sigma removes waste and makes the work culture and environment more efficient while reducing costs. This approach is required in South Sudan for the best water supply management.

Areas Requiring Changes in Water Kiosk Projects Under Lean Six Sigma Approach

The research was conducted by using relevant qualitative and quantitative data. Firsthand observations of the managers, workers, and vendors were obtained with the help of interviews and questionnaires. While doing so, it was identified that various areas could be upgraded for the removal of waste and efficiency of performance. As South Sudan recently got its independence and the developmental projects were handled mainly by the North, the infrastructure development plan has various flaws and deficiencies. If proper urban planning is applied to the infrastructure development in South Sudan, the water supply and management evolution will become much more manageable. This is one of the primary reasons causing issues in water supply and management. Similarly, another important kind of waste happening is the allocation of the same task to multiple water organizations. MWRI and SSUWC are both handling the urban water supply, which are a big waste of resources and a significant sign of mismanagement. It is not only a waste of economic resources but also of human resources. Removing such waste from the water management system of South Sudan is vital to make the processes smooth. The inventory is not being distributed efficiently as well. SSUWC can only keep 20 percent of the water bills received from the consumers, and the rest is sent to the federal government. The use of innovative technologies and techniques can further ease and pace up the development progress in South Sudan.

Sometimes, when multiple people are appointed to do the same job, it can be messed up. The job descriptions should be specified. Instead of wasting human resources, a single task should be appointed to a single individual. This kind of management is essential in a new country. It cannot bear the financial burdens, making it important that the resources are used wisely and efficiently.

Methodological Choice: Simple Mixed Method

The simple mixed method has been chosen for this study, as it required both qualitative and quantitative data collection/analysis (Cohen et al. 2017). The qualitative data was collected from prior studies and historical sources. The quantitative data was collected through questionnaires and organizational and government reports regarding the water issue in South Sudan.

Strategy: Action Research

This study was designed as action research as it aims to find solutions to the prevalent issues by interpreting the available data.

By using this strategy, the researcher is able to understand the issue in depth. Action research is also very effective because, along with finding solutions to the main issues, it identifies smaller flaws and waste that can be removed to increase the efficiency of the operation or process at hand (McNiff 2013). The action research divides the study into different steps, like background knowledge, identification of issues, planning, and implementation. This way, the required results can be practically achieved, and the overall culture of a country or organization can be changed. This strategy greatly influenced the accuracy of this study.

Process of Action Research

The action research follows a four-step process used in this research as described below.

- **a.) Context and Purpose.** The process of understanding the research triggers the genesis of action research. It is a pre-step that defines the context and purpose of the research. In this stage, it is also determined what effects the research will bring to the environment or culture it is applied to.
- **b.) Constructing.** In this phase, all stakeholders and people influencing the operations discuss the issues faced. The strategy for pre-planning and pre-action is also developed in this phase.
- **c.) Planning Action.** The data collected and arranged in the context and purpose and constructing phases are utilized

in the planning stage to design the action. The development of a complete action plan and number of actions required to solve the issues are also decided in this phase. In this stage, the initial steps required for the actions to be implemented are also planned and taken.

d.) Evaluating Action. After the planning stage is finalized, the actions are taken, and the results are recorded. The results can be either desired or undesired. The results are then compared with the constructing stage for identifying gaps. In the evaluation process, it is determined whether there was an issue in the planning or implementation of the action. Once the error is identified, it can be corrected before updating the design of the action plan. Here, one cycle is completed, and the next cycle starts. The evaluation phase of each cycle provides feedback to the constructing stage of the next cycle. The cycles continue until the required results are achieved.

Time Zone: Longitudinal

The time zone chosen for this study is longitudinal, as it is expected that it would require more than two cycles to find the right action plan for solving the water issue in South Sudan. The actions must be wisely designed, implemented, and evaluated to meet the desired objectives.

Techniques and Procedures

Research Methodology

To collect qualitative data related to the topic of this research, secondary sources were used, including books, journals, and articles. For the collection of quantitative data, official records of organizations, interviews, and questionnaires were used. This is a creative research with the objective of finding solutions and creating new opportunities (Collins 2010). The questionnaires and interviews were both conducted to enhance the authenticity of the research. The intention behind collecting both kinds of data was to find answers to the research questions and meet the aims of this research.

Data Analysis

The data for planning and implementing water projects with successful results were selected for analysis. The water situation in South Sudan was compared to the successful water projects to find the right combination for the newly independent country. Before the data analysis process, unreliable and invalid data were excluded. The reliability and validity of data were tested on three levels: first based on the source of the data, then on the grounds of the reason and the time of data collection, and finally according to the relevance of the data.

Data collection and analysis are major elements of any research, without which research cannot progress. The process in which the relevant data is identified, reviewed, and collected is known as data collection. The next process is analyzing this collected data, which includes testing the extent to which this data is reliable, valid, and relevant. This allows the researcher to be well-informed and prepared to conclude the research accurately. Ethical norms are applied in the process of data analysis for this research (Wiles 2012).

Identified Primary Stakeholders

Successful implementation of the actions is only possible if the primary stakeholders agree on the plan and implementation of the actions. The action plan and its implementation should be such that it should be acceptable to the stakeholders while meeting the objectives (Vijaya 2016). Prior studies examined for this research have revealed that the implementation of the actions has failed because the stakeholder synchronization factor was ignored. The stakeholder analysis for the South Sudan water kiosk projects is shown below.

Stakeholders for water issues in South Sudan
Citizens of South Sudan
African Development Bank
USAID
Government bodies for water supply and management
Private/community vendors
Water organizations
Project managers

It must be noted that government bodies, private vendors, water organizations, international NGOs, and project managers are all primary stakeholders making efforts to solve the water issue in South

Sudan. However, all these stakeholders are not synchronized, a major cause of previous projects' failure. It has also been observed that the number of stakeholders is too high, and the authority is unequally distributed among the stakeholders, causing waste in the operations. It is essential the stakeholders are compatible with each other and the action plan is such that all stakeholders are working in a unified manner.

Identification of Flaws and Waste in the Water Projects

During this study, it was revealed that there are certain areas where the project vendors and managers lack efficient operation. MWRI and SSUWC, along with some other local bodies, are all handling the water supply in the urban areas. This is not only a financial waste, but it is also causing a waste of human resources and management. When multiple organizations are monitoring a single task, it causes management issues that immediately reduce efficiency. Then other bodies are handling the rural water supply, working with different policies and budgets. The number of stakeholders in the South Sudan water issue has become so high that it is causing problems in the successful implementation of strategies. There are some areas where improvement is required to reduce waste and flaws. These areas, also major components of project management, are discussed below.

Targeted People

The study sample was a small section of the population. The demographics of this study involved the residents of village/rural areas in which the schemes were initiated, a delegate from the implementing agency, officials of the government water department, and community heads/leaders from the communities that hold several water projects. Therefore, this determined the exact intended number of participants the researcher had plotted in the zones where water schemes were initiated and the related agencies were involved in implementing and supervising these schemes.

Sampling Method

The sampling method involved selecting individuals to be included in the analysis as participants. The study ensured the chosen people were a good representation of the residents. The trial forms were developed depending on the role in the group, such as a leader of an executing agency or a member of the population involved. The researcher used a purposeful sampling technique and stratified sample methods. Purposeful sampling techniques were used in the case of representatives of institutions and water department officials. This was done because the researcher had to select some officials to participate in the study. According to Patton (2002, 230), the logic and power of purposeful sampling lie in selecting information-rich cases for in-depth study. Information-rich cases are those from which one can learn a great deal about issues of central importance to the purpose of the inquiry, thus the term "purposeful sampling." He further explained that studying information-rich cases yields insights and in-depth understanding rather than empirical generalizations. Stratified sampling was used for the selection of community members. This was done because the researcher needed to organize participants into manageable groups suitable for data collection. The researcher chose to collect data through focused group discussions to decrease data collection time because gathering information among public associates as a one-on-one conversation with this cluster would require much time and effort (both time and capital). These sampling systems confined the researcher's prospects of being prejudiced when selecting the study participants, as each interviewee had an equal chance of responding. The inhabitants were diverse, and each category would, therefore, have its survey/focused discussion guide. The research would reflect on these respondents as they have firsthand interaction with the social enterprise and could provide insight into the risk factors that affect the project.

Data Collection Instrument

The research would gather both qualitative and quantitative data using surveys and a focused group discussion guide. For questionnaires, by conducting a one-on-one survey (interview), the researcher used a semi-structured study to amass data. Semi-structured surveys are useful because they permit subjects to provide in-depth perceptions of a parameter. On the other hand, a yes or no response is restricted and would not require further details to be obtained by the researcher. The surveys were divided into different sections based on the defined study goals. The first segment was aimed at obtaining the participant's demographic data and explaining the background of the research. The second portion had inquiries covering different fields, and the last segment was to thank the subjects for their time and ensure the safety and privacy of the data shared (Creswell and Miller 2000).

Data Analysis Methods

This research project used a descriptive data analysis method after collecting information to make conclusions and recommendations. The analysis focused on interpreting collected data and determining patterns that emerged from the data (Creswell and Miller 2000). Since this research project was conducted using interviews recorded on audiotapes, the first part of the analysis involved coding this information by transcribing it into written text. When analyzing the data, the researchers transcribed all the recorded voice data into documents and clustered them into patterns for interpretation. The copied data were entered into a computer software program (Adobe Audition) and used pattern-matching logic to identify those that relate to each other and those that differ (Creswell and Miller 2000).

Computer software used for data analysis in this study was the

Statistical Packages for Social Sciences (SPSS) program, which allowed forming statistics for interpretation. SPSS was used when determining the mode and pattern of findings from participants regarding the risks of managing water kiosk projects (Creswell and Miller 2000). For example, the researcher determined the risk factor most mentioned during the interview as a challenge in managing water projects. From this data, analysis entailed comparing risk factors less mentioned during the interviews. Descriptive statistics were used in determining the mode and making conclusions. This is important because a factor mentioned by all participants should be prioritized over a risk factor indicated by just one participant in the study (Creswell and Miller 2000).

Lastly, data analysis for this result involved identifying themes from the findings, which were then used in making conclusions and recommendations. In this research project, the themes emerging from the study included significant risks affecting the management of water kiosks and how they consequently affect the entire community due to their need for freshwater (Creswell and Miller 2000). In essence, data analysis focused on interpreting the collected data to form findings and conclusions. The interpretation involved identifying patterns, themes, and other issues that may arise from the study.

Validity and Reliability

A pilot evaluation was the first aspect of validation and reliability. Piloting requires pre-testing the appropriate tools to assess the quality of the sample with a minor section of subjects. To avoid exposure, subjects who participated in the experimental phase were not used again for the main analysis. Besides, the study piloting process helped standardize the instruments before they were used to collect data. Only two subjects from each target cluster were chosen for the pilot study. The research methods were then checked based on the observations in the trial study (Golafshani 2003).

In essence, the researcher focused on getting accurate data to disperse any concerns about validity and credibility. The study ensured it got the right data that answers the research question and contributes to the literature (Golafshani 2003). This was achieved by avoiding any bias that could lead to falsified information, especially when forming findings and conclusions from the collected data. The researcher also considered the applicability of the information to a broader audience as one way of facilitating validity. The focus was on whether information from the research conducted in rural South Sudan could be applied in other regions worldwide (Golafshani 2003). In essence, the researcher was looking to inform policymakers and strategists who want to sustain the performance of community water projects.

Instruments of Validity

The validity of a tool assists in assessing the severity of the results. The institution and the reviewing body's research and test-measuring experts were used to verify each tool used in the analysis. The devices were thus circulated for review and feedback to the assigned professor and a few other colleagues. In terms of content and overall validity, they helped evaluate the instruments. These individuals' opinions helped to ensure the remarks in respective questionnaires have the information needed in a precise and candid manner, according to the study's aim.

Data Collection Process

The LIGS University issued an identity card (ID) and an introduction letter used to obtain the necessary study approvals from the appropriate research and innovation ruling body. The researcher traveled to the study area after receiving a permit and obtained permission from the resident government and made appointments with the relevant subjects before gathering the required data. The researcher used a one-on-one discussion (interview) strategy for some topics and a focused group discussion with members of the public from the concerned zones. Since focused group discussions involve two facilitators, the researcher recruited skilled staff who served as the notetakers. These appointments were also helped in performing more analysis and clarifying unclear points. Researchers ensured they got the right sample population for data collection, which was fit to answer the research question (Golafshani 2003). Participants had a crucial role in determining the validity and reliability of collected information based on their cooperation in data collection. Therefore, the researcher ensured that individuals only participated voluntarily and that their consent was included as a part of the project.

Limitations of the Research Study

The qualitative study used both primary and secondary data. This method is not without its shortcomings or limitations. Notably, one of these limitations is the time taken to conduct the study. According to Creswell (2014), Janesick (2011), and Miles et al. (2014), this type of study consumes a lot of time and requires an intensive labor force. Getting the parents' consent to interview children about the time when they fetch water could sometimes affect the researcher. The researcher conducted a pilot study and explained why it was important to get children's views regarding water projects. No disclosure of important information could also affect data collection and interpretation. In this case, the researcher used both open and closed-ended questions supplemented by probed questions to get the right data from respondents. Prior to collecting the data, the researcher visited some project sites and made some inquiries about the water kiosk projects.

Other limitations included the return of some interview

questionnaires, which were delivered in order to adhere to the social distance policy of COVID-19. The researcher provided enough information to the respondents on the purpose and use of the research data. Also, the researcher faced challenges in the data collection due to the total lockdown of South Sudan, following the pandemic of coronavirus, or COVID-19. The lockdown policy affected data collection because some respondents could not be accessed, while other respondents could not freely provide information due to the imposed condition of social distancing, use of face masks, avoidance of handshakes, and, above all, stay-at-home regulations.

Lastly, all the kiosks studied are developed by NGOs. The projects were transferred to the local community, whose capacity to document and manage information was limited. Hence basic important project information was not easily accessible.

Research Discussion, Analysis, and Findings

Introduction

South Sudan is a newly-born infant state that saw two fierce civil wars, one before and one right after the independence. The first war was fought for an extended period of time, from 1983 to 2005, until South Sudan achieved independence. The other started right after the independence and was fought between the government and rebels. As the state did not have many economic deposits, the wars took the state into a financial deficit. North Sudan was mainly in control of all water and sanitation works before the independence. South Sudan was in a relatively better situation at the time of independence, in regards to water supply with the help of NGOs. However, the water and sanitation development of South Sudan has been greatly affected due to the second war and economic crisis. That is one of the reasons why only 40 percent of the population in South Sudan has access to water for basic needs.

Another flaw identified in this study is that different institutions are handling urban and rural water supplies. Not only that, the urban water and sanitation management is being monitored by multiple organizations, which is causing a waste of resources and affecting the output flow badly.

Data Management and Representation

Two sets of instruments were used to collect primary data from the respondents: questionnaires and interviews. Questionnaires were used for data collection from managers, water organizations, and water vendors or users to identify the issues in community water management in South Sudan. A total of eighty questionnaires were distributed: twenty for managers, twenty for water organizations, and forty for water vendors. Out of the questionnaires distributed, for the managers, there was a 90 percent response rate, 85 percent response rate from water organizations, and 100 percent response rate from the vendors. Also, twenty interviews were conducted to collect qualitative data from the respondents. The data representation and respondent details are provided below for each question individually.

Findings

Findings from the Water Managers

The number of respondents in this questionnaire was eighteen in total, with fourteen males and four females. The percentage of male respondents was 77.8 percent, while the females accounted for 22.2 percent.

Figure 13. Respondents in the Water Manager Questionnaire

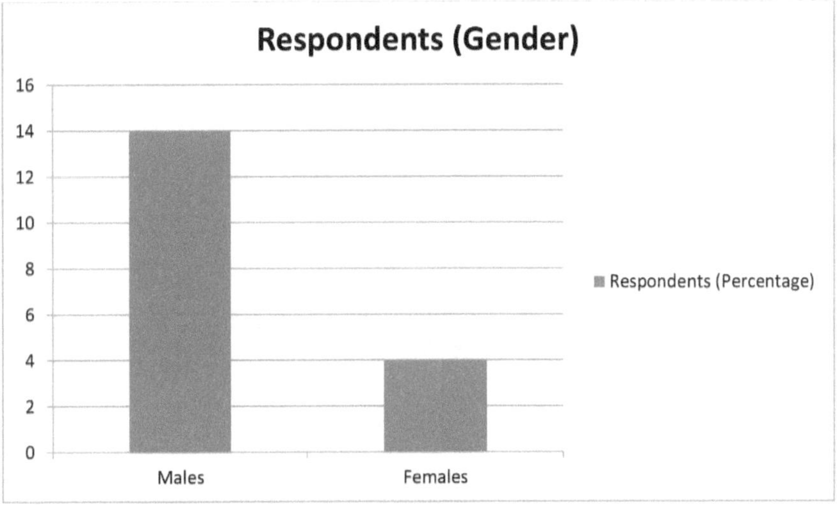

Respondents (Gender)

Gender	Respondents (Percentage)
Males	14
Females	4

Note: graphic provided by the author

The respondents were selected so each of the eighteen respondents has a different way of life, including student, sultan (chief), government officer, businessperson, and several others. Each respondent belongs to a different profession, so the insights collected are versatile and cover a significant population sample. The following chart describes it in detail.

Figure 14. Occupation of the Respondents

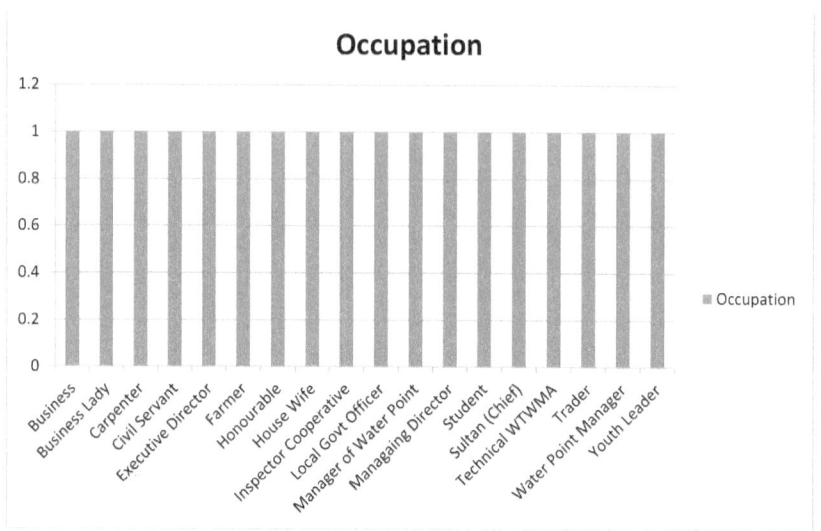

Note: graphic provided by the author

The education level of the respondents also showed a variance represented in the following chart.

Figure 15. Education Level of Respondents

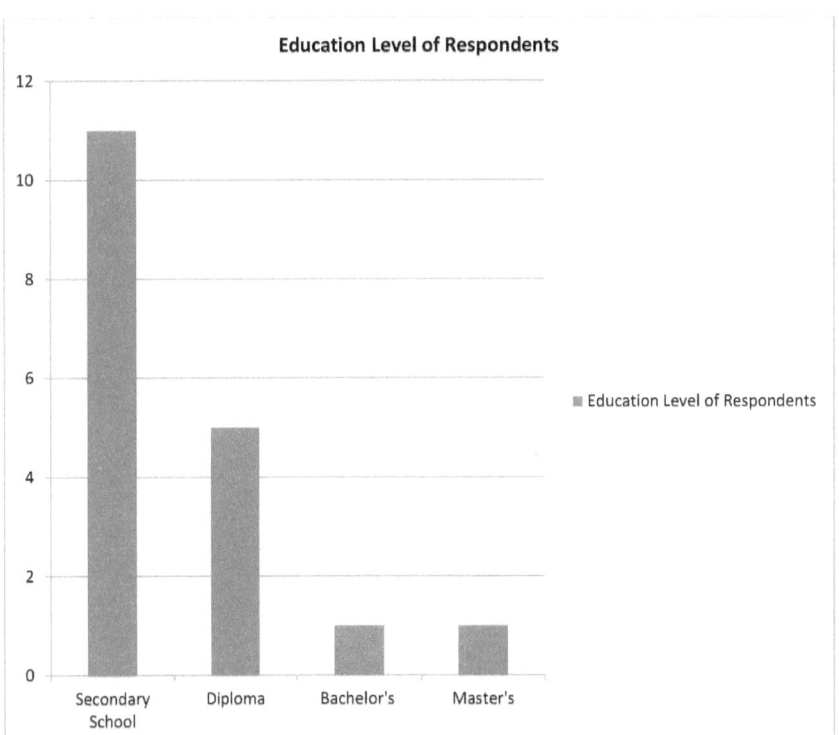

Note: graphic provided by the author

Another important factor kept in mind was choosing managers with different experience levels, which are shown in the following chart.

Figure 16. Experience Level of Respondents

Experience Level of Respondents

Experience	Count
1-5 years	3
5-10 years	8
10-15 years	2
15-20 years	3
Above 20 yrs	2

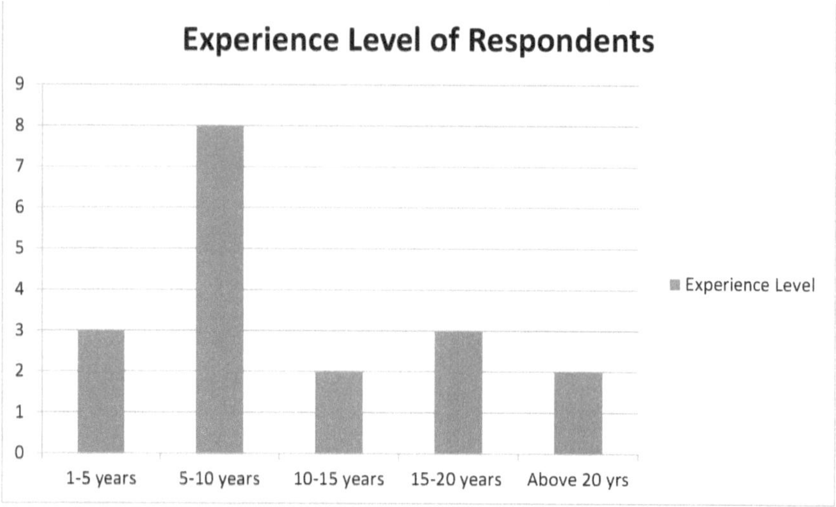

Note: graphic provided by the author

It can be observed from the above chart that respondents from experience level varying from one year to twenty years and more were included to get the response of the experienced minds as well as the young minds. The next factor upon which the respondents for this questionnaire were analyzed is the age group. The age group of the respondents was also selected in such a way that versatile feedback was obtained. The age group of the respondents is shown below.

Figure 17. Respondents' Age Group

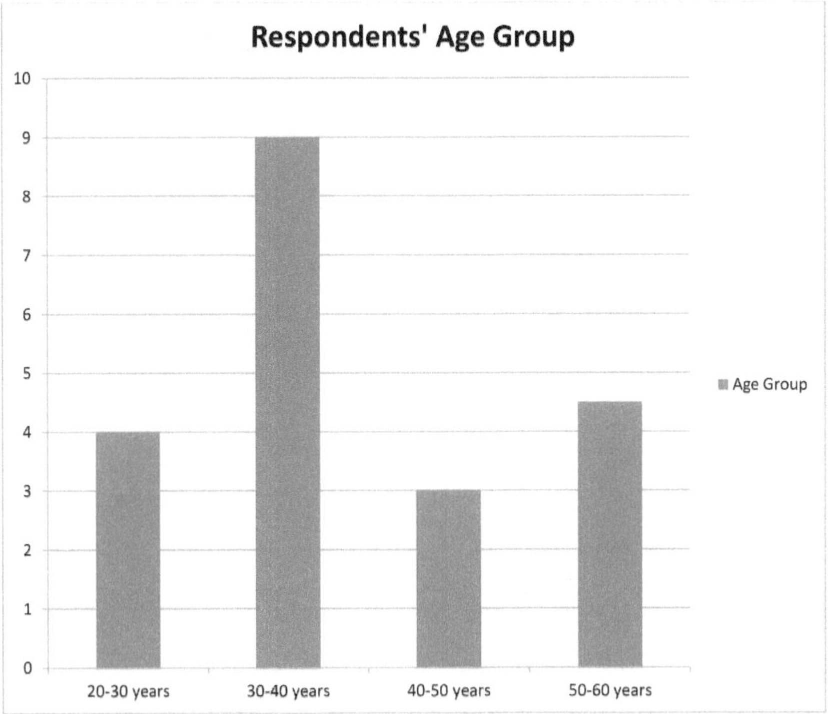
Respondents' Age Group
10
9
8
7
6
5
4
3
2
1
0
20-30 years
30-40 years
40-50 years
50-60 years
Age Group

Note: graphic provided by the author

All the respondents of the manager's questionnaire were asked whether they have managed any water project before, and all responded affirmatively. In the next question, they were asked about the number of projects they have managed and the duration of those projects. The following illustration will provide the details.

Figure 18. Number of Projects and Duration

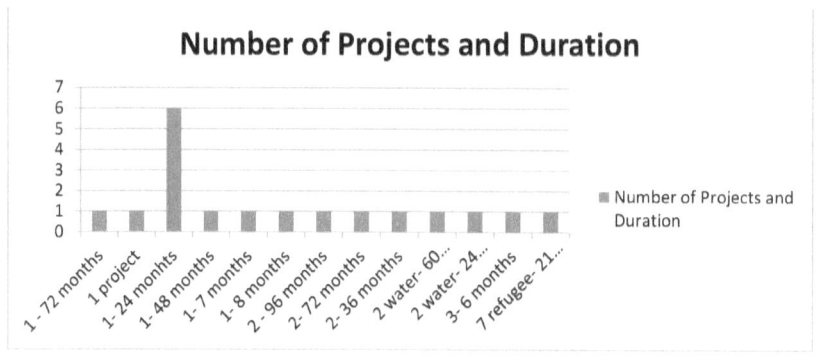

Note: graphic provided by the author

The next question was about the planning and implementation, and the respondents were asked to provide their opinions on whether they believe it was done in the right way or not. Their responses are illustrated in the following chart.

Figure 19. Planning and Implementation of the Projects

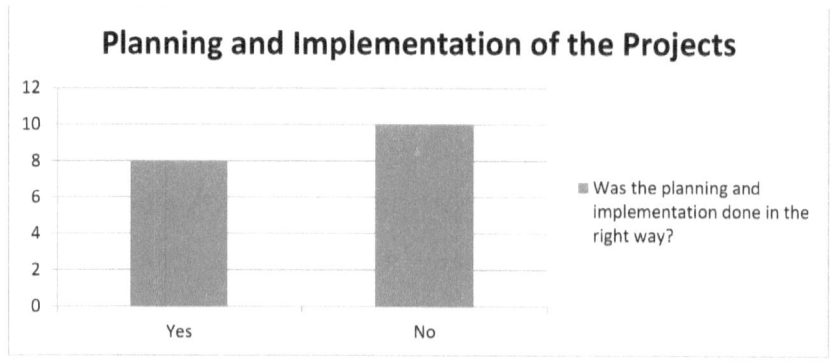

Note: graphic provided by the author

It is clear from this chart that the majority of the managers felt that the planning and implementation of the projects were not done as it was supposed to be. 44.4 percent of respondents said the planning and implementation was done well. It was important to know why they thought the implementation was unsuccessful, the next question of the questionnaire. The response is presented below.

Figure 20. Reasons Behind Unsuccessful Project Implementation

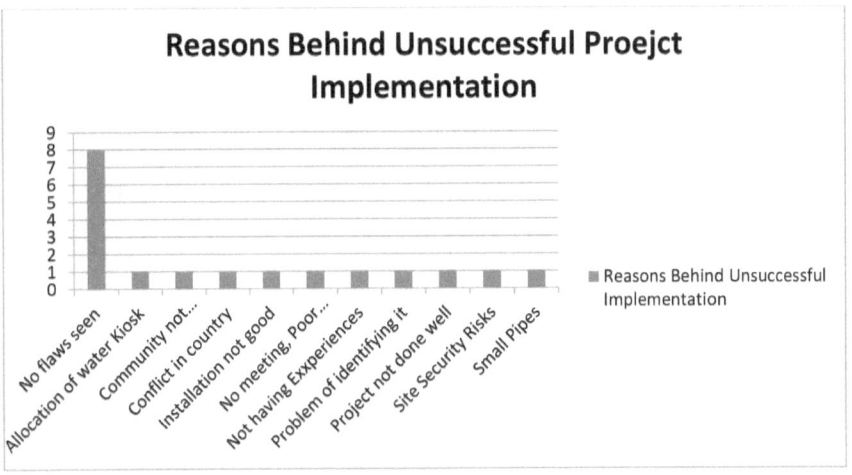

Note: graphic provided by the author

44.4 percent of respondents think there are no issues in implementing the projects. The remaining 55.5 percent provided a reason, and all of these reasons impact the low efficiency of water supply and management in South Sudan. Allocation of the water kiosks is a factor that can cause problems in the implementation of the water projects if not done in the right way. It was also observed that 5.5 percent of respondents stated that the community, a primary stakeholder in water projects, was not involved, which caused failure in the past. The conflict within the country is also an important factor causing the projects to slow down. The equipment-related issues have also been mentioned.

The communication level was pointed out by one respondent who claimed that no meetings were held and the project was managed poorly. The respondents also mentioned security risks, lack of experience, and smaller pipe usage. The next question was designed to identify the stage of the project implementation where the managers thought they faced challenges. 38.8 percent of respondents felt they faced challenges during the selection of the site, 27.7 percent of them thought it was during the construction of the water kiosk, 16.6 percent found challenges in the operations, and 16.6 percent considered the challenges present in the monitoring/evaluation of the projects. A graphical illustration of the responses to this question is provided below.

Figure 21. Stage of Projects Posing Challenges to Water Managers

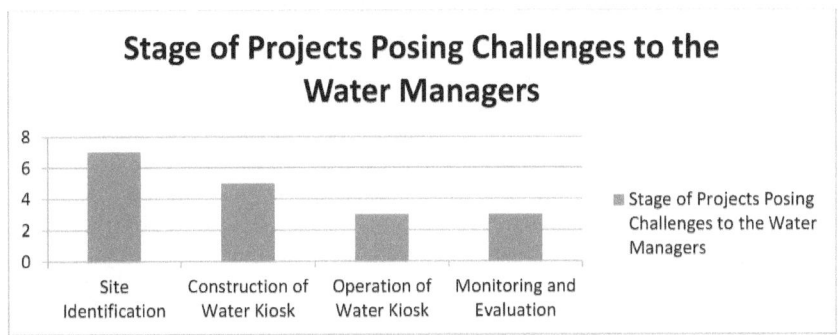

Note: graphic provided by the author

After identifying the different stages when water managers face challenges, it was also important to determine the nature of these challenges. In response to this question, 50 percent of respondents thought the cultural issues were a big challenge, 11.1 percent believed that lack of good communication was the challenge, 33.3 percent felt it was due to financial constraints, and 5.5 percent chose "other" as the answer. The responses are shown in the following chart.

Figure 22. Challenges in Project Management

```
Challenges in Project Management
10
 9  ■
 8  ■
 7  ■
 6  ■                    ■
 5  ■                    ■         ■ Challenges in Project
 4  ■                    ■             Management
 3  ■                    ■
 2  ■         ■          ■
 1  ■         ■          ■         ■
 0
   Cultural Issues  Communication  Financial  Others
```

Note: graphic provided by the author

The next question aimed to determine the manager's view, if they believed there are risks related to water projects in South Sudan. All 100 percent of respondents believed this was true and responded that the projects were risky. The common answer was that risks were there, but it was important to determine the nature of these risks. The next question, with four parts, prompted their response over different kinds of risks. Part A of question 9 asked the respondents about the legal risks, and all responded that they did not exist.

Figure 23. Legal Risks

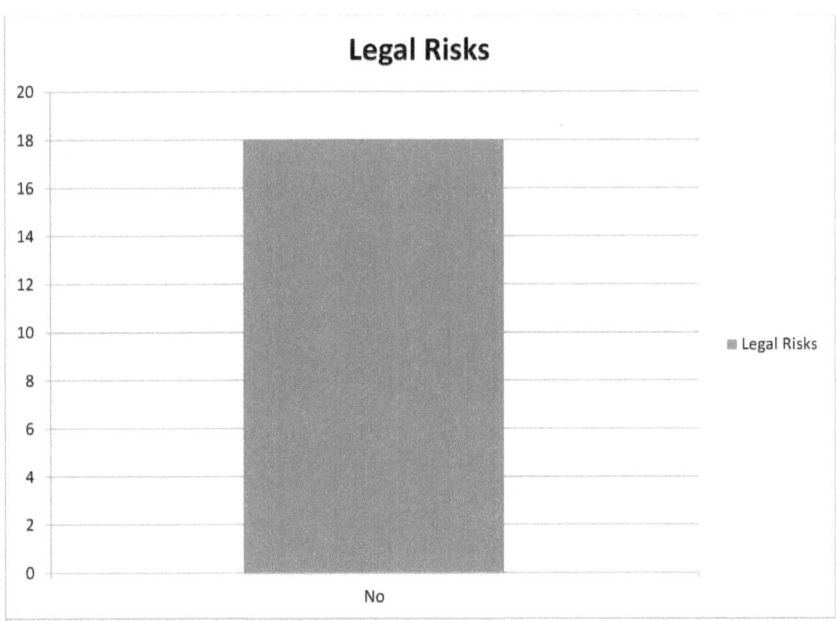

Note: graphic provided by the author

Part B inquired about the political risks related to water projects in South Sudan. In response to this part, 22.2 percent of respondents thought political risks were involved, while 77.7 percent thought political risks were not involved.

Figure 24. Political Risks

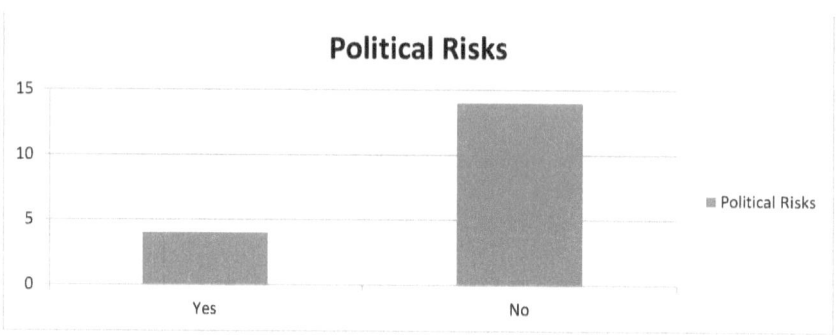

Note: graphic provided by the author

Part C aimed to get the opinion of the managers regarding the financial risks. 83.3 percent of the managers replied that financial risks were present, while 16.6 percent thought there were no financial risks.

Figure 25. Financial Risks

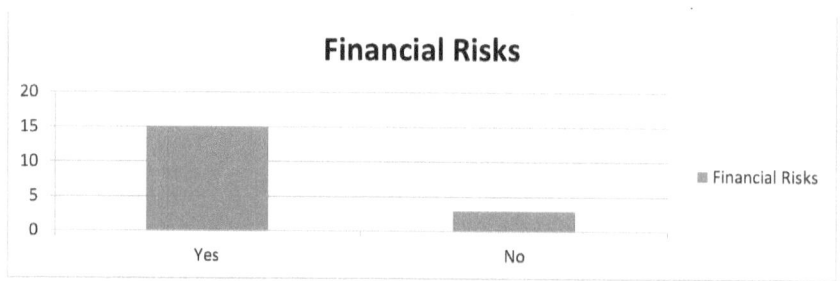

Note: graphic provided by the author

Part D of this question was about the technical risks involved in the water kiosk projects of South Sudan. 55.5 percent of respondents thought the technical risks were a reality, while the other 44.4 percent did not think so.

Figure 26. Technical Risks

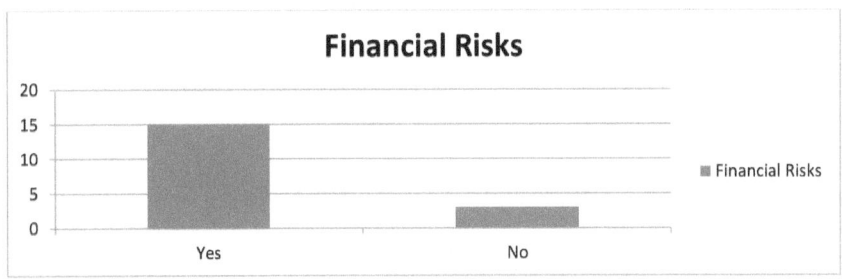

Note: graphic provided by the author

Similarly, part E of this question was about the security risk related to water projects. 44.4 percent of respondents were prone to believe that security risks were present, while the other 55.5 percent thought there were no security risks to the projects.

Figure 27. Security Risks

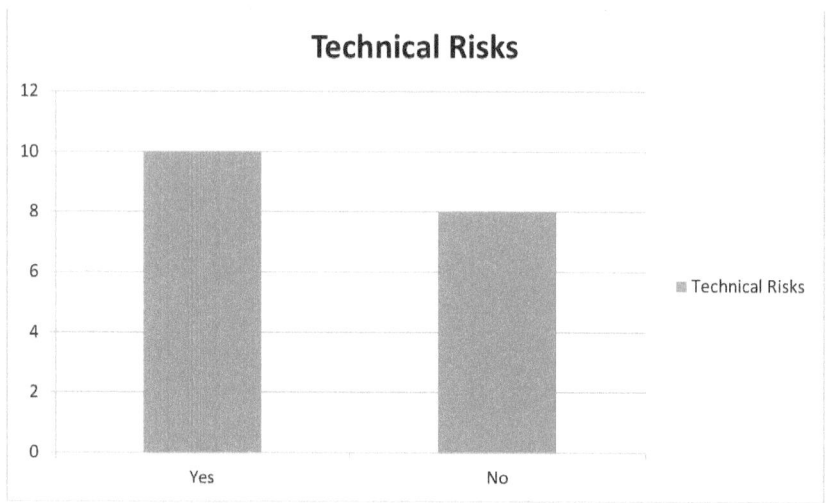

Note: graphic provided by the author

Managerial risks are also part of the projects, which was asked in the next part of this question. 55.5 percent of managers felt the risk related to management was there, while 44.4 percent believed otherwise.

Figure 28. Management Risks

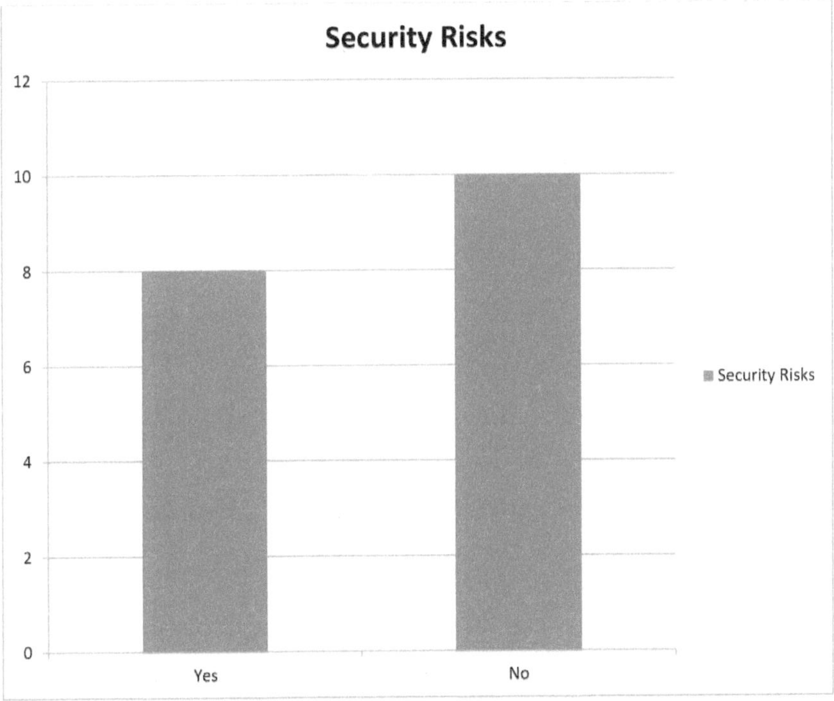

Security Risks

Category	Count
Yes	8
No	10

Note: graphic provided by the author

After getting the managerial response regarding different kinds of risks, they were asked to provide solutions to avoid or mitigate these risks. All the managers provided a unique answer, and all of these solutions can actually help the water supply and management projects of South Sudan. The solutions provided by the managers are illustrated in the chart below.

Figure 29. Solutions to Reduce Risks

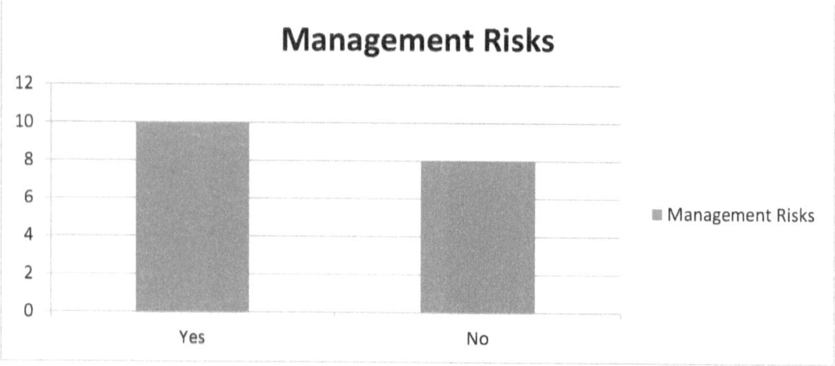

Management Risks

(Bar chart: Yes ≈ 10, No ≈ 8; legend: Management Risks)

Note: graphic provided by the author

It can be seen in the above chart that all the respondents provided a unique solution for the removal of risks. All these options are correct in regard to the corresponding risk. The experience of these managers and the solutions provided by them should be included in the constructing phase of developing a new action plan for the water kiosk projects in South Sudan. It can be seen that the solutions are provided for all kinds of risks defined in the previous questions.

Next, the respondents were asked to provide information regarding project guidelines and policies. 88.8 percent of respondents believe that guidelines and policies were provided by the community, but 11.1 percent of respondents did not agree. The response can be observed in the following illustration.

Figure 30. Presence of Community Guidelines/Policies

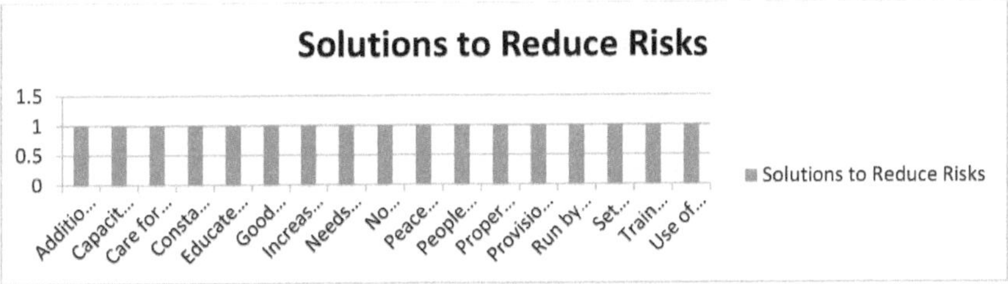

Solutions to Reduce Risks

[Bar chart showing categories: Additio..., Capacit..., Care for..., Consta..., Educate..., Good..., Increas..., Needs..., No..., Peace..., People..., Proper..., Provisio..., Run by..., Set..., Train..., Use of... — all with values around 1, legend: Solutions to Reduce Risks]

Note: graphic provided by the author

In the next question, the managers were asked which areas, according to them, are following the community guidelines and policies. 16.6 percent of respondents did not answer, while the rest provided specific answers. Only 11.1 percent of respondents gave a common response related to the fetching period. After analyzing the responses to this question, it was observed that the guideline is mostly implemented for the bill payments due to the consumers plus the number of households defined per water point. The rest of the areas lack guidelines on their implementation. The detailed responses are provided below.

Figure 31. Implementation of Guidelines/Policies

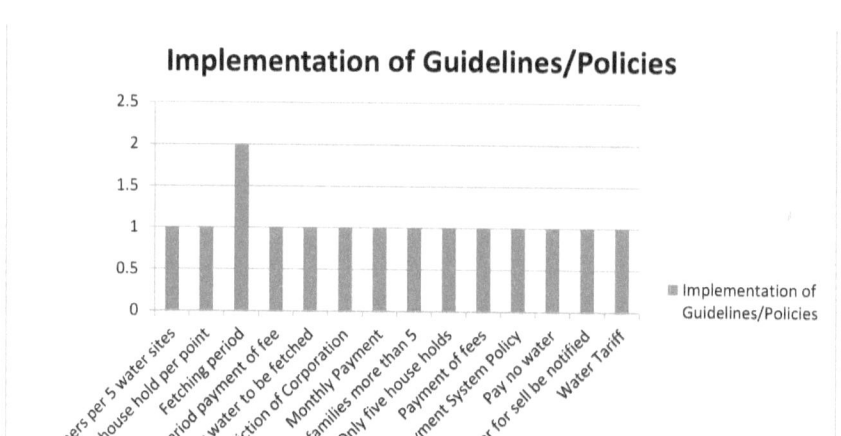

Note: graphic provided by the author

After that, the respondents were asked to give their opinion that if a certain project started now, would it be in working condition even after ten years? Only 22.2 percent of the respondents believed it would be continued, while 77.7 percent believed it would not.

Figure 32. Still Working After Ten Years?

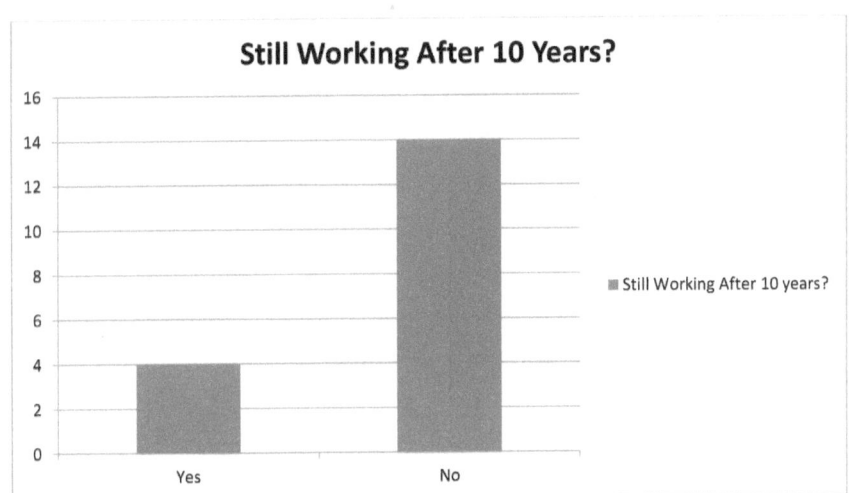

Note: graphic provided by the author

They were then asked to provide the reasons behind the fact it would not be working after ten years. 27.7 percent of the respondents did not answer this question. The rest, 72.3 percent, provided their responses. Out of this figure, 5.5 percent of respondents believe the constant breakdown was a big issue that might cause the project to close down. 11.1 percent of respondents believe the financial constraint was the reason why the project would discontinue. 55.9 percent of the respondents believe the cause of the project shutdown would be poor management. Theft, user negligence, and lack of technical support were also some responses. The detailed response analysis for this question is as follows.

Figure 33. Reasons for Project Shutdown

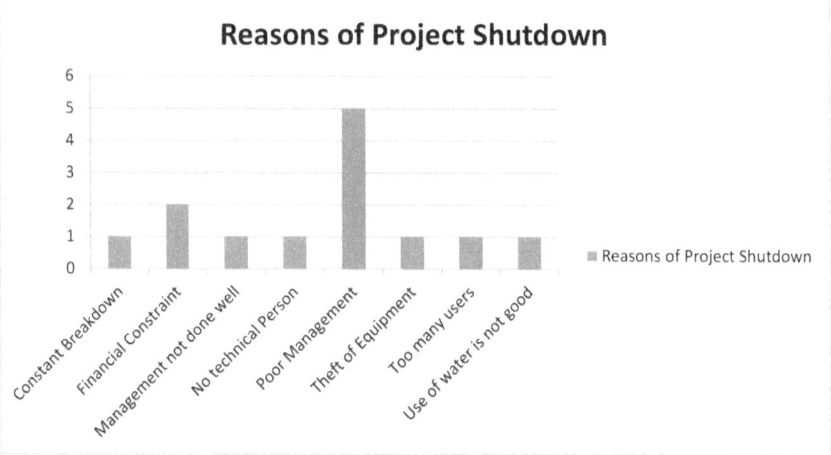

Reasons of Project Shutdown
6
5
4
3
2
1
0
Constant Breakdown
Financial Constraint
Management not done well
No technical Person
Poor Management
Theft of Equipment
Too many users
Use of water is not good
Reasons of Project Shutdown

Note: graphic provided by the author

In the next question, the respondents were asked to provide their feedback regarding the areas that can effectively provide sustainability to water projects in South Sudan. First, they were asked whether the harmonized policies could effectively increase the sustainability of the projects. 27.7 percent of respondents responded positively, while the other 72.2 percent answered negatively.

Figure 34. Project Sustainability—Harmonized Policies

Project Sustainability - Harmonized Policies
15
10
5
0
Yes
No
Proect Sustainability - Harmonized Policies

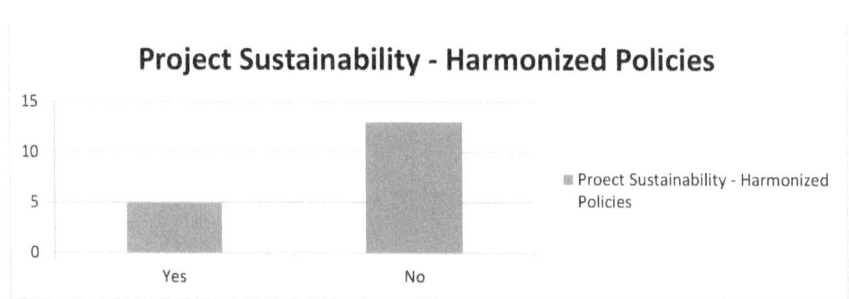

Note: graphic provided by the author

The next part regarded capacity training and whether it can help improve the sustainability of projects. 83.3 percent of managers believe capacity training could be helpful, while the rest of them, 16.6 percent, did not believe that.

Figure 35. Project Sustainability—Capacity Training

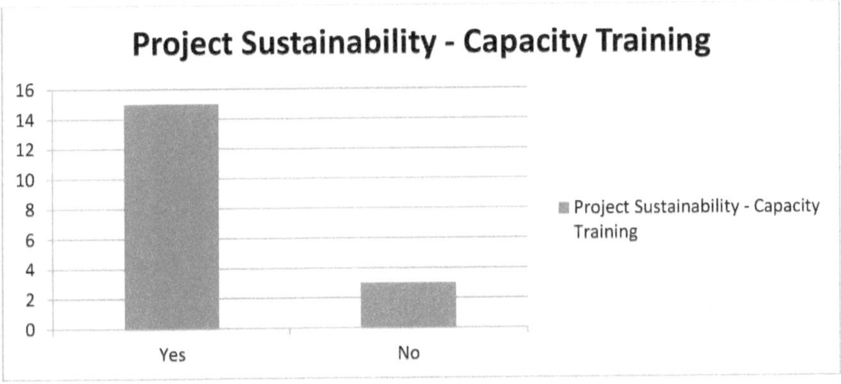

Note: graphic provided by the author

In the next part of this question, they were questioned about the water rates and whether they could help make the projects sustainable. 38.8 percent of respondents replied affirmatively, while 61.1 percent of respondents answered negatively.

Figure 36. Project Sustainability—Water Rates

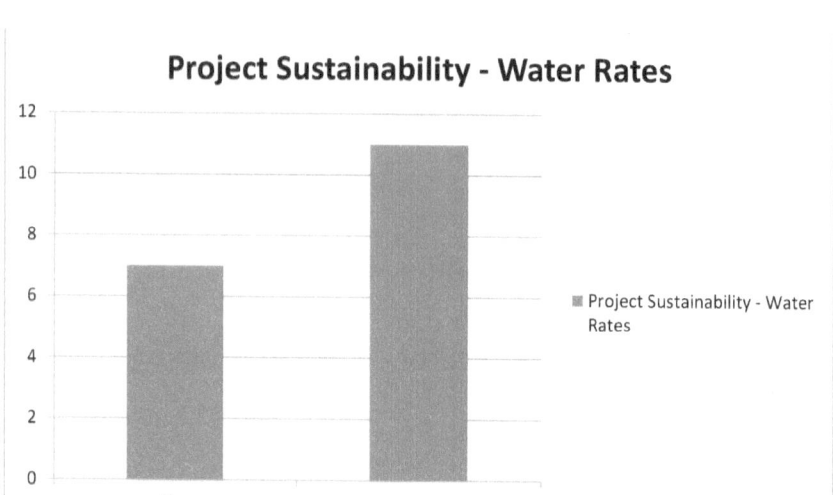

Note: graphic provided by the author

The managers were then asked about customer policy control and how effective they could be in achieving the sustainability of water projects. Only 11.1 percent of managers believe it could help, while the rest were not prone to believe it influences the sustainability of water projects.

Figure 37. Project Sustainability—Customer Control

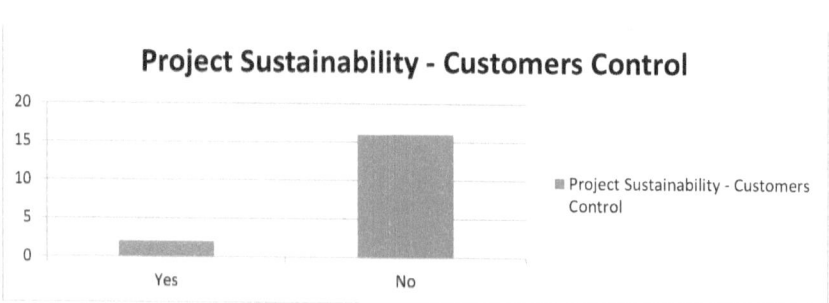

Note: graphic provided by the author

The next part inquired about regular maintenance and its influence on the sustainability of the water projects. 66.6 percent positive responses were registered, and only 33.3 percent were negative ones.

Figure 38. Project Sustainability—Regular Maintenance

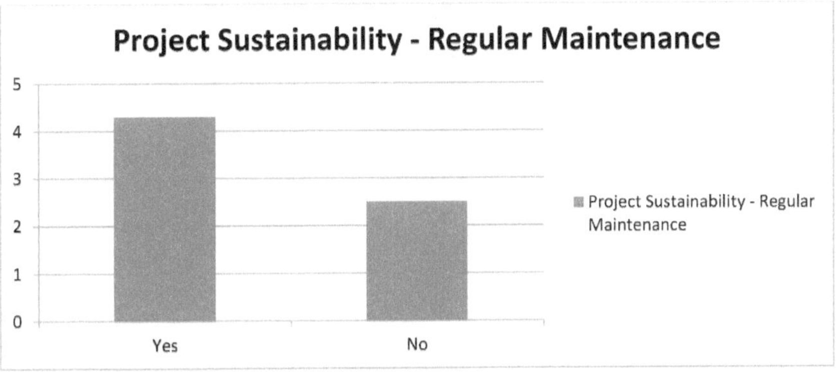

Note: graphic provided by the author

Part F of question 15 of the manager's questionnaire regarded the control of fetching time and its effects on the sustainability of the water kiosk projects. Only 22.2 percent of respondents believe it could be effective, while 77.7 percent did not consider it an influential factor.

Figure 39. Project Sustainability—Control Fetching Time

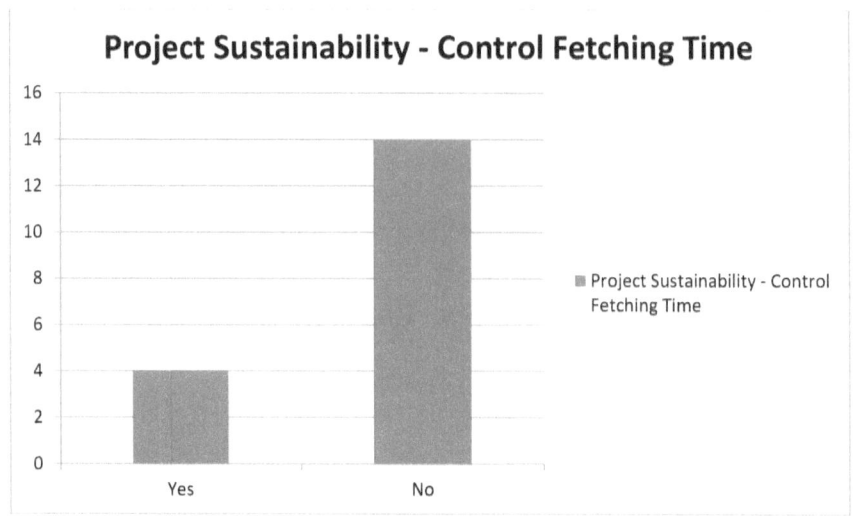

Note: graphic provided by the author

The final part of this question was about the influence of periodic monitoring on the sustainability of a project. 33.3 percent of respondents believe it was influential, but 66.6 percent believe it was not.

Figure 40. Project Sustainability—Periodic Monitoring

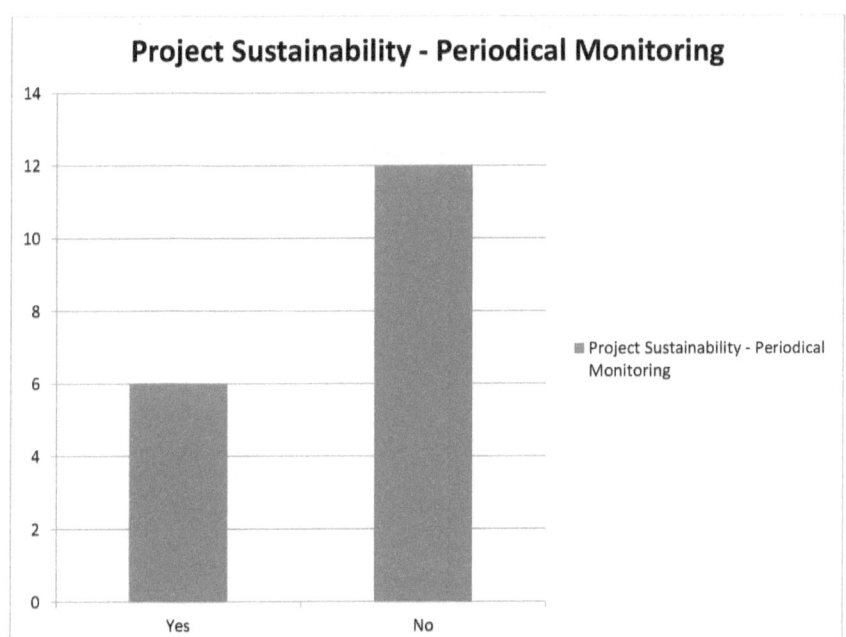

Note: graphic provided by the author

The next question was about getting recommendations from the water managers on improving the water projects in South Sudan. The first part of this question inquired whether the involvement of the stakeholders helped in improving the process. 77.7 percent of managers believed it would improve the water projects, while 22.2 percent thought otherwise.

Figure 41. Improvement of Water Projects—Stakeholder's Involvement

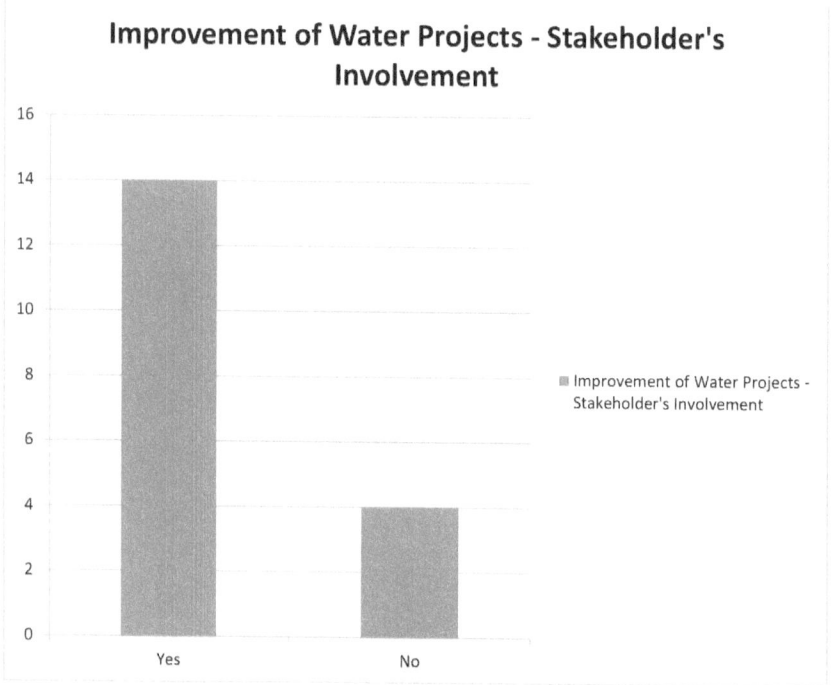

Note: graphic provided by the author

The next part was about the effect of regular meetings on improving water projects. 44.4 percent of respondents believed it could help improve water projects, but 55.5 percent did not agree with this.

Figure 42. Improvement of Water Projects—Regular Meetings

[Bar chart titled "Improvement of Water Projects - Regular Meetings" showing Yes = 8, No = 10]

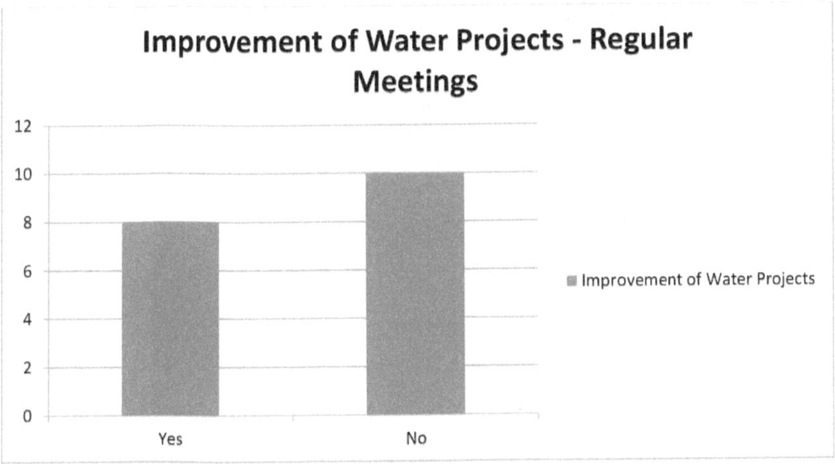

Note: graphic provided by the author

In the next part of this question, the relationship between timely supervision and improvement of water projects was explored. 44.4 percent believed it could be effective, while 55.5 percent believed it was not influential in improving water projects.

Figure 43. Improvement of Water Projects—Timely Supervision

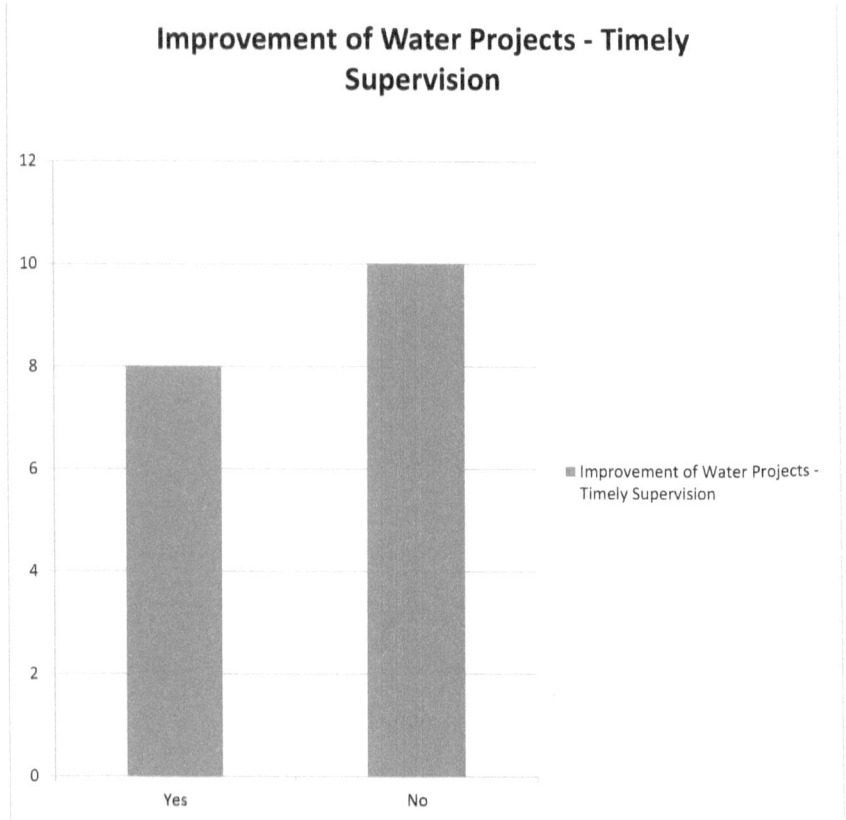

Note: graphic provided by the author

Alternative funding can be a factor in the improvement of water projects. The managers were asked about this factor. 55.5 percent of the respondents answered affirmatively, while the other 44.4 percent negatively.

Figure 44. Improvement of Water Projects—Alternative Funding

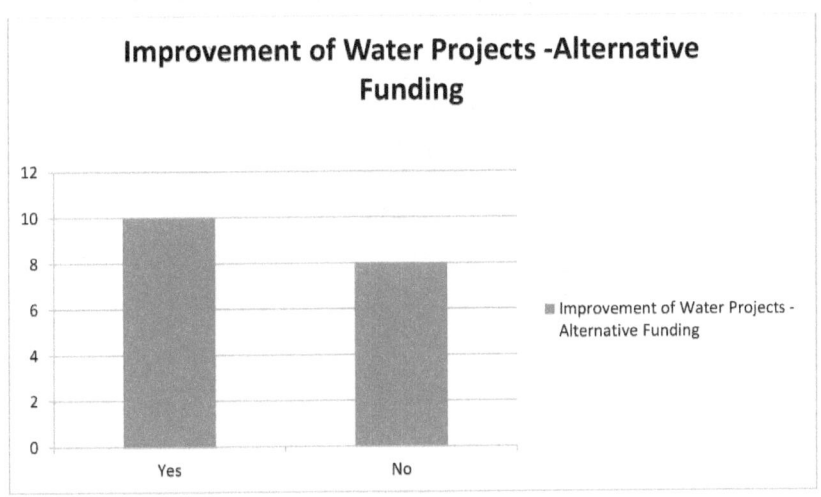

Note: graphic provided by the author

Lastly, the respondents were asked to provide recommendations not already present in the questionnaire. Fifteen respondents did not answer this, while the rest of the responses are shown in the following chart.

Figure 45. Improvement of Water Projects—Others

Improvement of Water Projects - Others

[Bar chart showing three bars all at approximately 1.0 for: Avail Spare Parts, Recall Contractor, Sustainable Water]

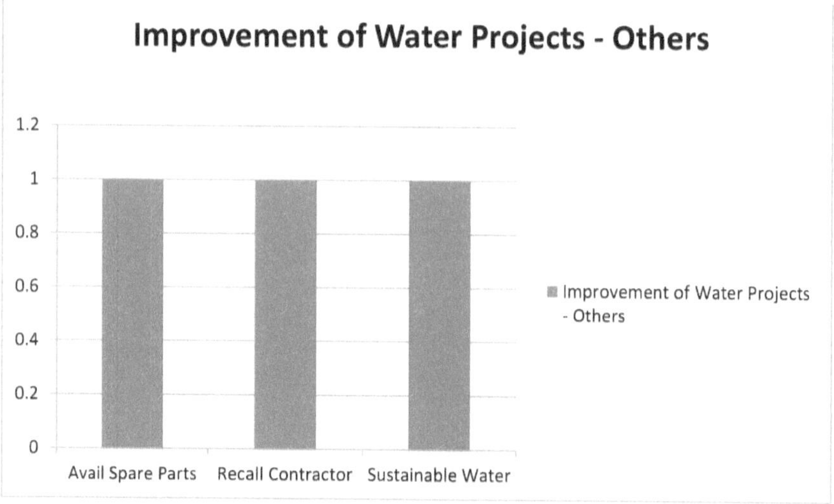

Note: graphic provided by the author

Findings from the Water Organizations

Along with collecting data from water managers, it was also important to obtain feedback from the water organizations regarding their experience and knowledge of the operations. For this purpose, the second questionnaire was designed with twelve questions and distributed to several water organizations working in South Sudan. Twenty questionnaires were distributed and filled by seventeen different individuals representing sixteen unique water organizations. Selected companies are currently working or have past experience with water projects in South Sudan. The following chart shows the percentage of respondents representing selected water organizations. The company name data is shown in the following chart.

Figure 46. Water Organizations

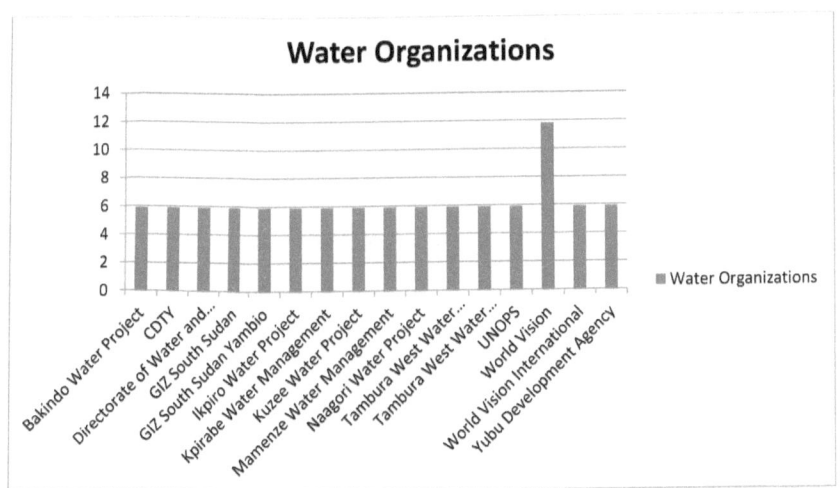

Note: graphic provided by the author

The next question regarded the number of years operating. The range of years operating is between one and seventy for these companies. The number of companies is located on the Y-axis, while the X-axis represents the years of operating for the corresponding companies. The detailed data are presented in the chart below.

Figure 47. Years of Operation

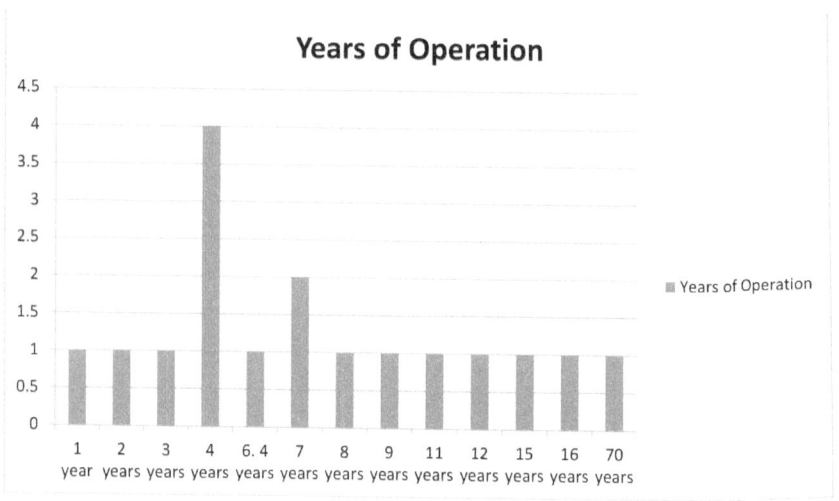

Note: graphic provided by the author

These versatile years of operation of different water organizations have enabled this research to simultaneously obtain data from young and experienced companies. The next question was about the number of employees working in these organizations. Again, the data analysis showed that all companies have a different number of employees, enabling the categorization of big or small companies. According to the data collected in this question, it was found that the biggest company in the respondent list has fifty-eight employees, and the smallest company has only one employee. The collection of information from the perspective of both larger and smaller companies strengthens the authenticity of this research. The Y-axis represents the number of companies, and the X-axis represents the number of employees. The details of the employment status of these organizations are illustrated in the following chart.

Figure 48. Number of Employees

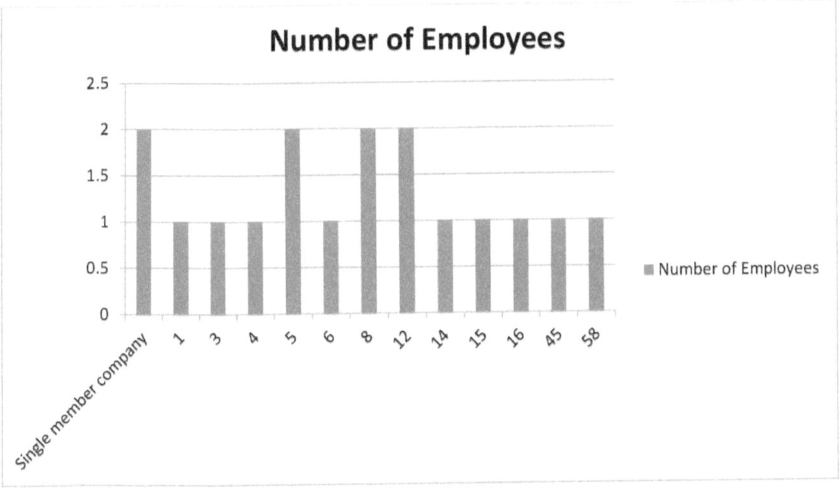

[Bar chart titled "Number of Employees" with y-axis ranging from 0 to 2.5. X-axis categories: Single member company (2), 1 (1), 3 (1), 4 (1), 5 (2), 6 (1), 8 (2), 12 (2), 14 (1), 15 (1), 16 (1), 45 (1), 58 (1).]

Note: graphic provided by the author

The next question asked the respondents about the number of operations their corresponding company manages. The selected companies handle between one and twenty-seven projects. The insights into the company handling twenty-seven projects greatly helped study the complexities and hindrances of water issues at a macro level. On the other hand, insights into the companies handling single projects assisted in interpreting and understanding the issue at a micro level. Six companies handle just one operation, four companies handle two water operations, three companies handle five operations, and others handle multiple projects. The data analysis of the number of operations of the selected companies is graphically presented below. The number of operations is shown on the X-axis, and the number of companies is shown on the Y-axis.

Figure 49. Number of Operations in Progress

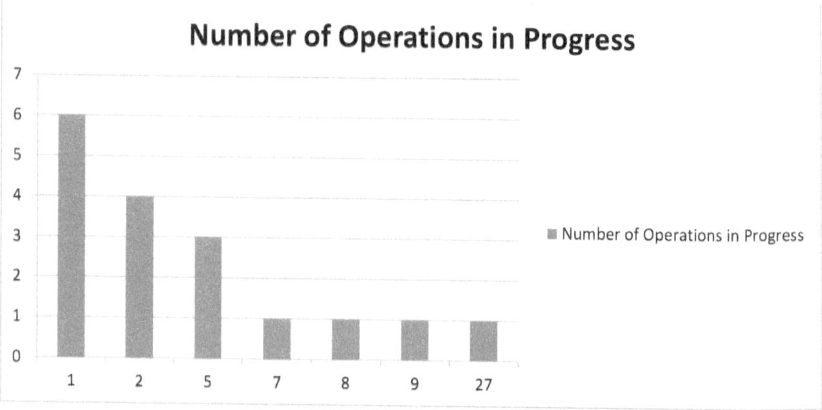

Number of Operations	Number of Companies
1	6
2	4
5	3
7	1
8	1
9	1
27	1

Note: graphic provided by the author

The next question inquired whether it is their first community project. Sixteen organizations responded it was not their first community project, and only one company claimed they had not done it before. Almost all the companies have experience managing a community water project.

The respondents were then invited to share the risks they had previously faced in such projects. Three respondents did not provide input. The other fourteen respondents provided feedback based on their previous experiences. It has been observed from the data collected that the cultural perception of water being free is a big challenge faced by South Sudan water projects. The risk of theft and equipment breakdown risk have also been identified. The political, managerial, and security risks have also been highlighted. Detailed data analysis of this feedback is shown in the following chart.

Figure 50. Previously Faced Risks

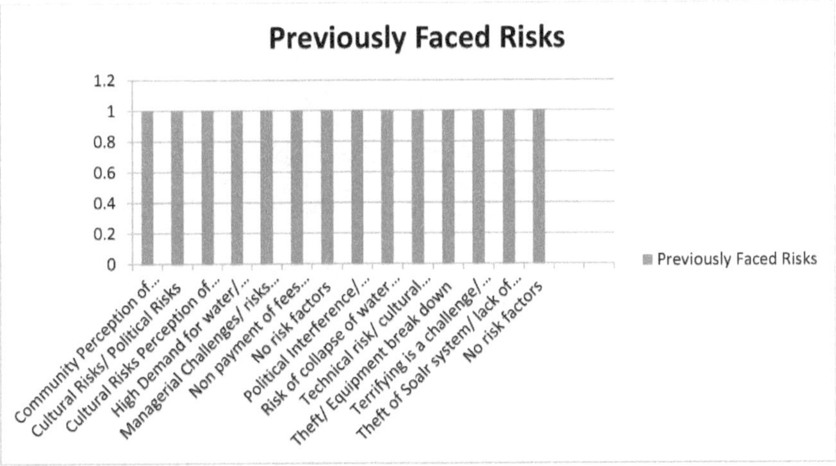

Note: graphic provided by the author

After that, the respondents were asked to explain how their respective companies mitigated these risks. Different solutions were mentioned. Conducting meetings and asking for funds from other organizations were provided as solutions. Some suggestions were made regarding authoritative adjustments, constant repairs, employment of security guards, and creation of awareness among the locals. These all are ways in which these companies have handled the risks and challenges. Five respondents did not give their input in this part of the question. Detailed data representation is given in the chart below.

Figure 51. Recommendations for Risk Handling

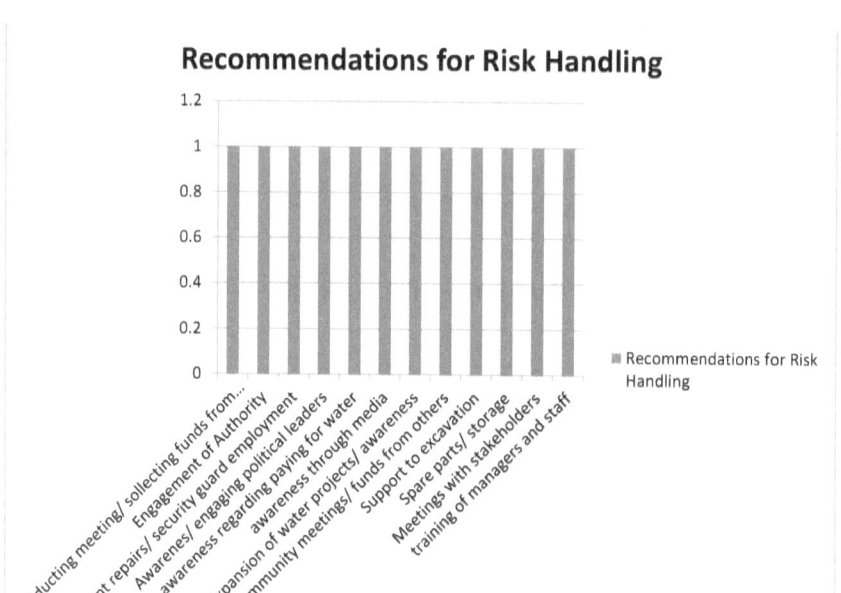

Note: graphic provided by the author

After that, the next question asked for their opinions about the community's involvement in the water projects. Six respondents believe that involving the community can help water projects in South Sudan succeed. 94.1 percent of the respondents believed the community should be involved. The data is presented graphically as follows.

Figure 52. Involvement of the Community

```
Involvement of the Community
Yes: 16
No: 1
```

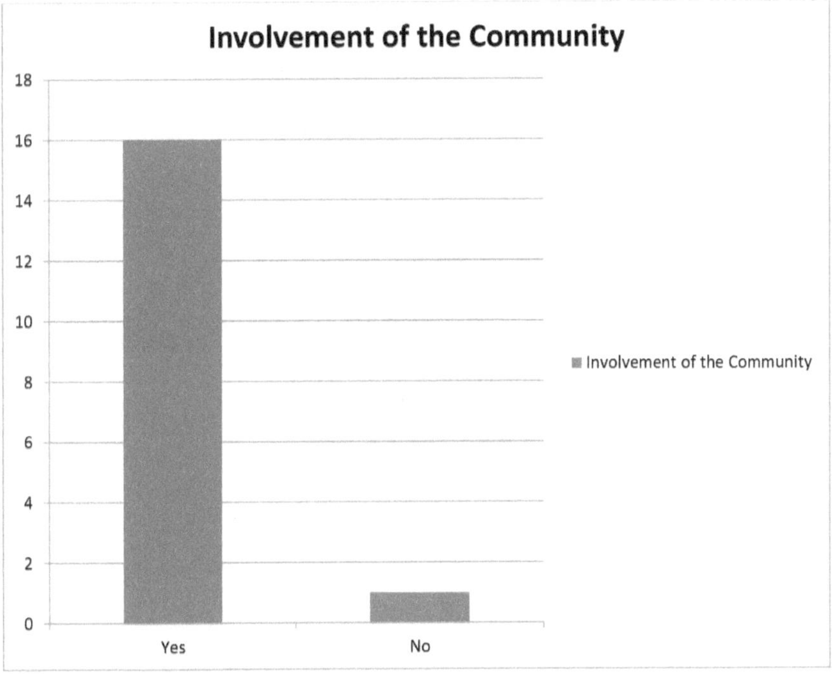

Note: graphic provided by the author

Part B of this question asked for solutions implemented by these companies to engage the community in water projects. Sixteen responses were collected for this part. The answers provided by the respondents are unique and random and are illustrated in the following chart.

Figure 53. Engaging the Community

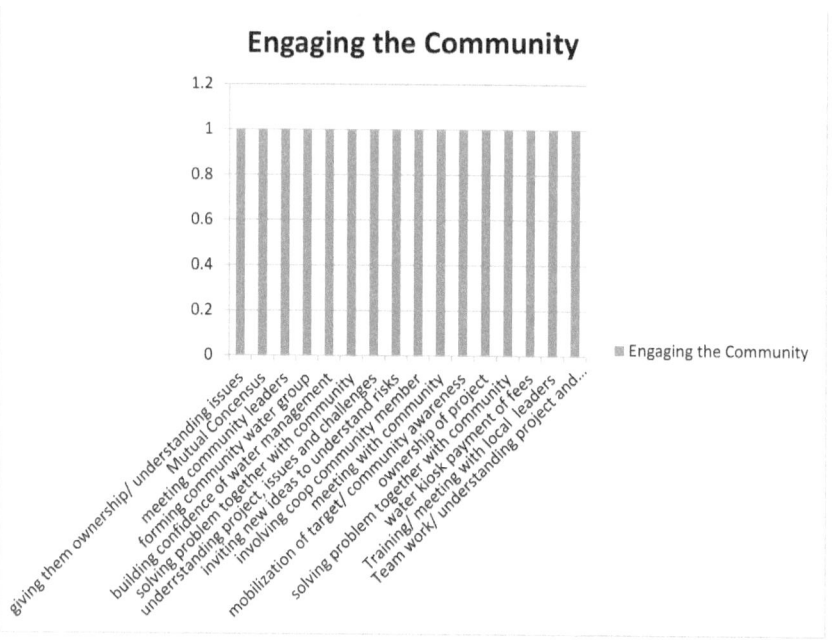

Note: graphic provided by the author

The next question aimed to get the respondents' input on whether the current water projects face any challenges. 100 percent of the respondents believe the risks were already present.

Figure 54. Water Projects Facing Challenges

```
Water Projects Facing Challenges
18
16
14
12
10
 8                              ■ Water Projects Facing Challenges
 6
 4
 2
 0
         Yes
```

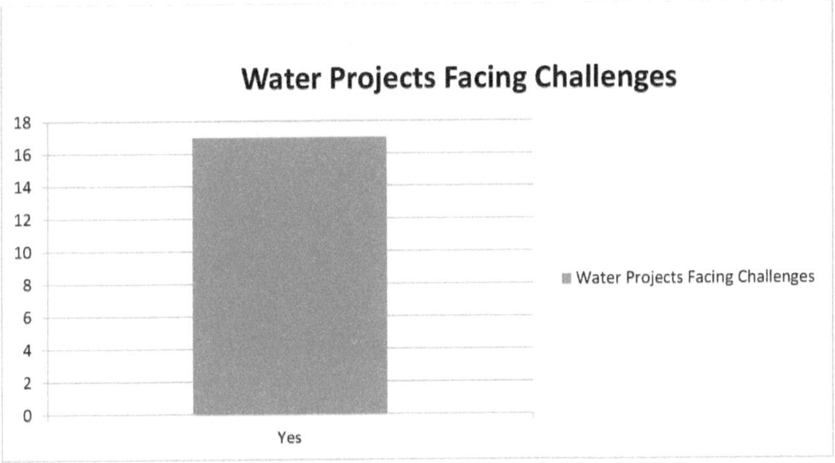

Note: graphic provided by the author

In the next part of this question, they were asked to elaborate on some of the challenges. 17.6 percent of the respondents believed the projects faced financial challenges. The rest of the respondents provided their own descriptions of the challenges. Most respondents were prone to believe that the financial and managerial challenges were the biggest ones causing hindrances to the success of water projects. Some respondents also highlighted a lack of spare parts and technical support. Security challenges have also been identified. Detailed data representation is provided in the chart.

Figure 55. Challenges of Water Projects

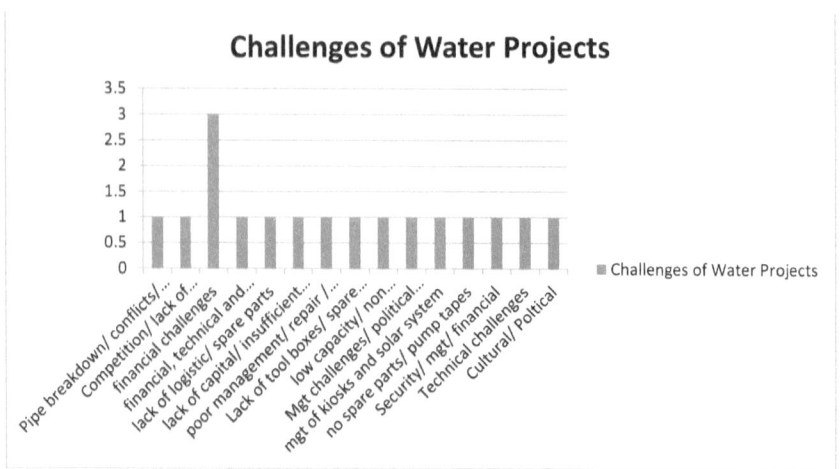

Note: graphic provided by the author

The next question was quite similar but focused only on the current challenges in the projects. The results were almost the same as described in the above chart. The companies were asked how they have mitigated the risks and challenges previously and how they are handling them currently. Interestingly, the feedback for these three questions was almost identical. It seems the challenges and risks present back then are still faced currently. Moreover, the approach and ways the companies use have not changed much as well. These water organizations are using the same approach and measures to mitigate these risks and challenges (Awadh and Saad 2013).

In accordance with this, the next question asked the respondents about the planning done to avoid the reoccurrence of these risks. 47.1 percent believed that planning is implemented, while 52.9 percent thought otherwise.

Figure 56. Planning to Stop Risks From Reoccurring

Planning to Stop Risks From Reoccurring
- Yes: 8
- No: 9

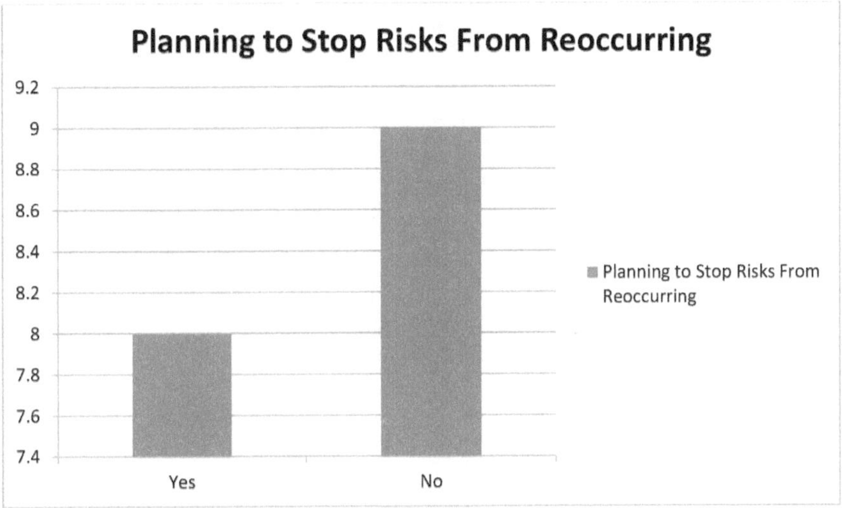

Note: graphic provided by the author

The next part of this question was about measures taken to prevent these risks from reoccurring. Nine respondents did not provide an answer to this question, and the rest provided measures via their specific companies. These measures include sanitation, punishments, and expansion of projects. The data collected for this part is graphically illustrated below.

Figure 57. Measures to Stop Risks From Reoccurring

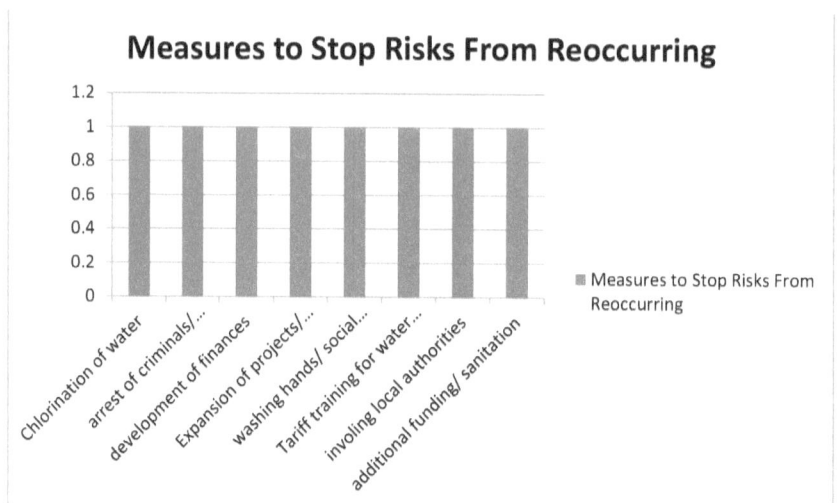

Note: graphic provided by the author

The questionnaire filled by respondents from different water organizations provided in-depth knowledge and insight regarding water projects' risks and challenges in South Sudan.

Findings from the Water Vendors

After collecting data from the water managers and organizations, the last stakeholder group was the water vendors. Forty respondents filled out this questionnaire, among which 53.7 percent were males and 46.3 percent were females.

Figure 58. Gender (Water Vendors)

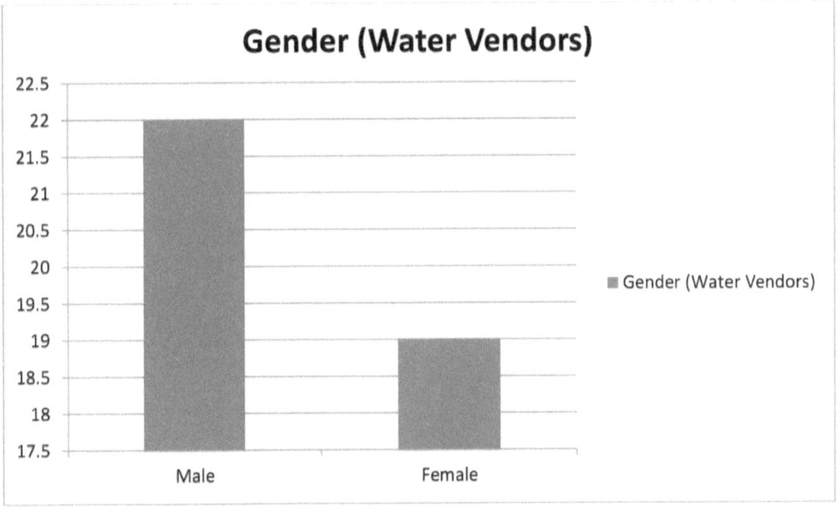

Note: graphic provided by the author

Every person in the community needs water, including the managers and operators. The respondents selected for this questionnaire belonged to different fields, which provided accurate and authentic data collection regarding the issue. The respondents' occupations are as versatile as farmers, businesspersons, civil servants, finance assistants, and many more. Most of the respondents have different occupations.

The next question regarded the educational level of the vendors. The data analysis of this question showed that 12.2 percent are graduates, 17.1 percent are diploma holders, and 70.7 percent have a secondary school education.

Figure 59. Education Level

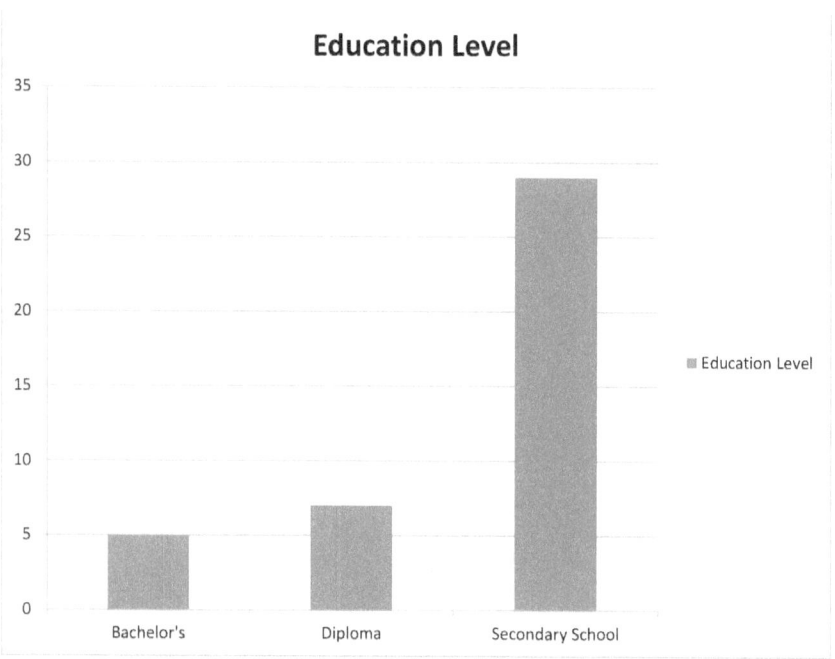

Note: graphic provided by the author

The next question regarded the years of experience of the respondents. 58.5 percent of respondents have one to five years of experience, 22 percent have five to ten years of experience, 9.8 percent have ten to fifteen years of experience, 2.4 percent have fifteen to twenty years of experience, and 7.3 percent have more than twenty years of experience.

Figure 60. Years of Experience

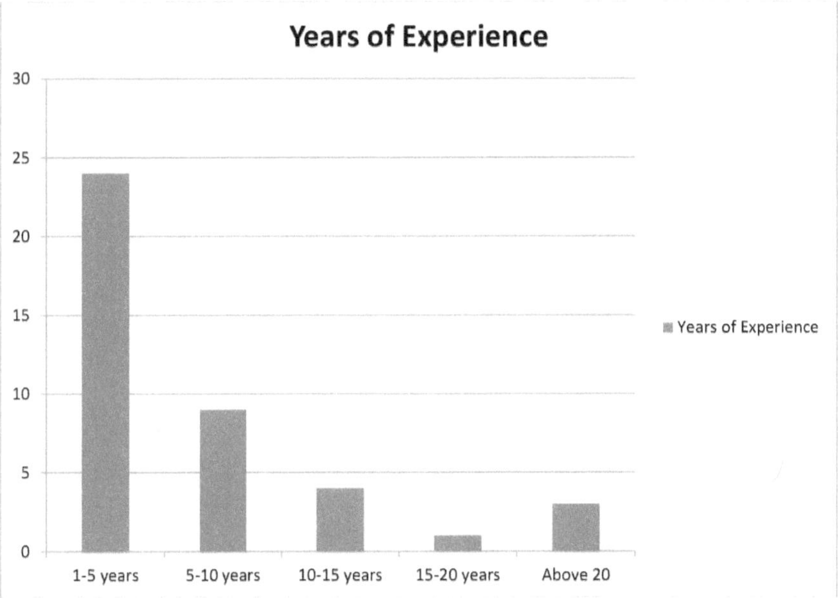

Years of Experience

Range	Count
1-5 years	24
5-10 years	9
10-15 years	4
15-20 years	1
Above 20	3

Note: graphic provided by the author

The next question was about the age group to understand the respondents' mental capacity and capability. 39 percent belong to the twenty to thirty years age group, 31.7 percent belong to the thirty to forty years age group, 22 percent are in the forty to fifty years age group, 2.4 percent are in the fifty to sixty years age group, and 2.4 percent belong to above sixty years age group.

Figure 61. Age Group

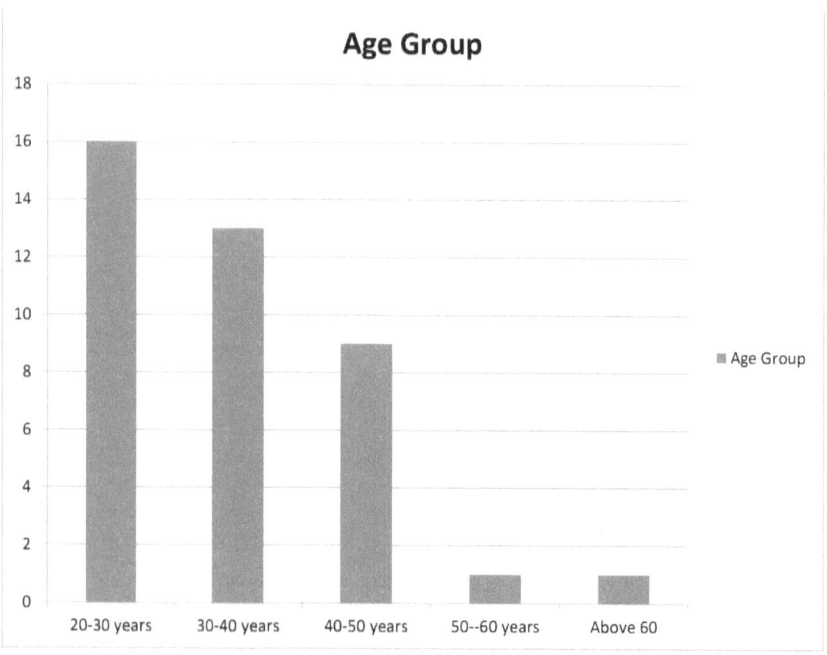

Note: graphic provided by the author

The next part of this question asked whether the respondents were residents of the village in question. 97.6 percent of respondents were residents of the village, and only 2.4 percent were not.

Figure 62. Residents or Not

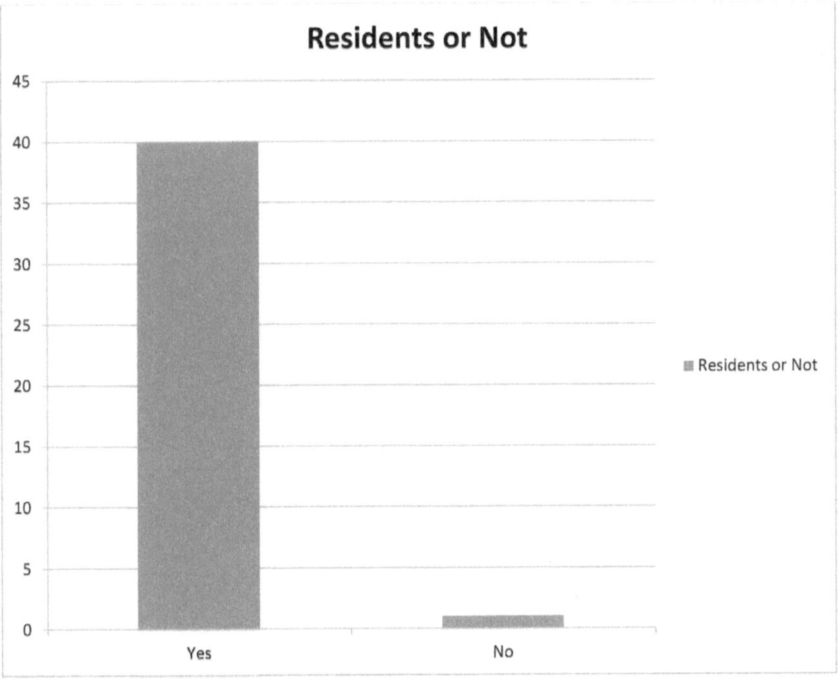

Note: graphic provided by the author

The number of people in a household affects water consumption. The respondents were asked to give information about the number of members in their household. 26.8 percent had one to five people in their household, 65.9 percent had six to ten, 4.9 percent had eleven to fifteen, and 2.4 percent had sixteen to twenty people living in their household. The data is illustrated in the following chart.

Figure 63. Number of People in Household

Number of People In Household

Category	Count
1-5 people	11
6-10 people	27
11-15 people	2
16-20 people	1

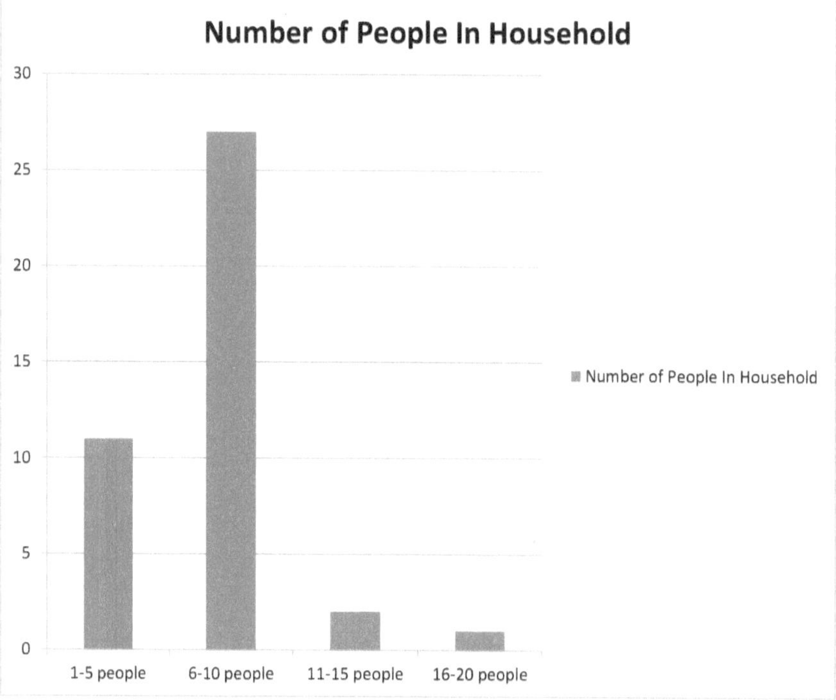

Note: graphic provided by the author

This was the end of the first question and all its parts. The second question asks the respondents about the primary water sources available in the village. 90.2 percent of respondents answered it was tap water, 2.4 percent said it was the river, and 7.3 percent mentioned boreholes/open wells. The X-axis shows the options, and the Y-axis represents the number of responses.

Figure 64. Source of Water

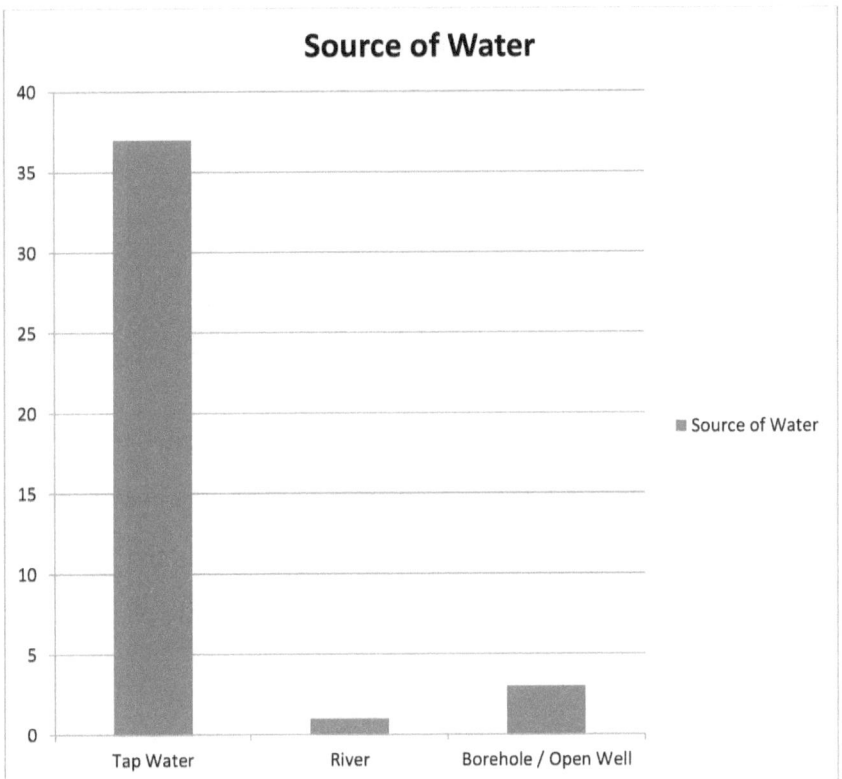

Note: graphic provided by the author

In the following question, the respondents were asked how many years they think the water project has served the community members. 63.4 percent of respondents believed that it had served for one to five years, 24.4 percent believed it to be five to ten years, 4.9 percent thought ten to fifteen years, and 2.4 percent selected fifteen to twenty years. The data is illustrated as follows.

Figure 65. Number of Years Water Project Has Served

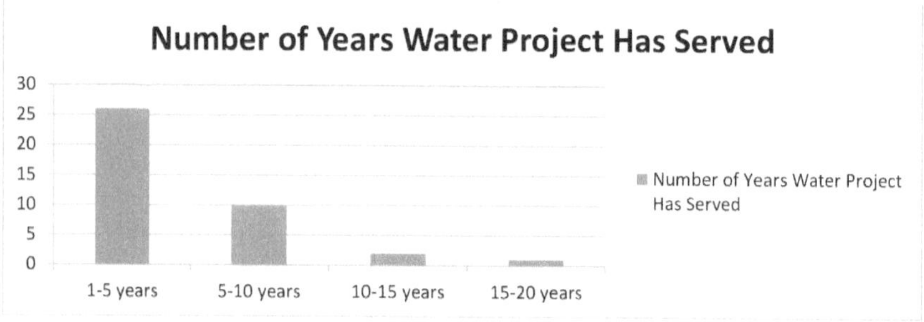

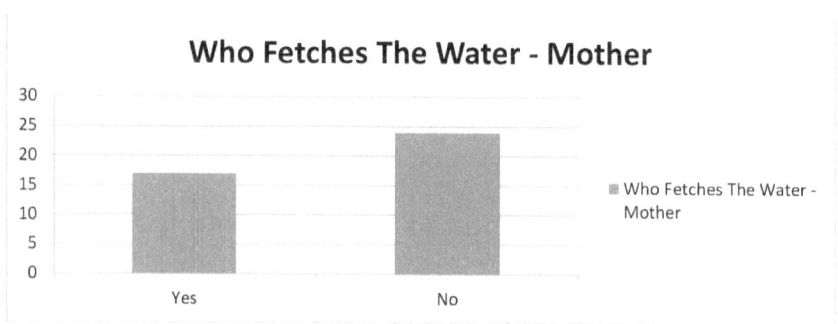

Note: graphic provided by the author

The fourth question was divided into several parts to collect the response regarding the person who collects the water in the household. The first option was the mother of the household. 41.5 percent of respondents responded affirmatively, while 58.5 percent negatively.

Figure 66. Who Fetches The Water—Mother

Note: graphic provided by the author

The next option was the father; only 2.4 percent of respondents replied the father fetched the water, and 97.6 percent responded negatively.

Figure 67. Who Fetches The Water—Father

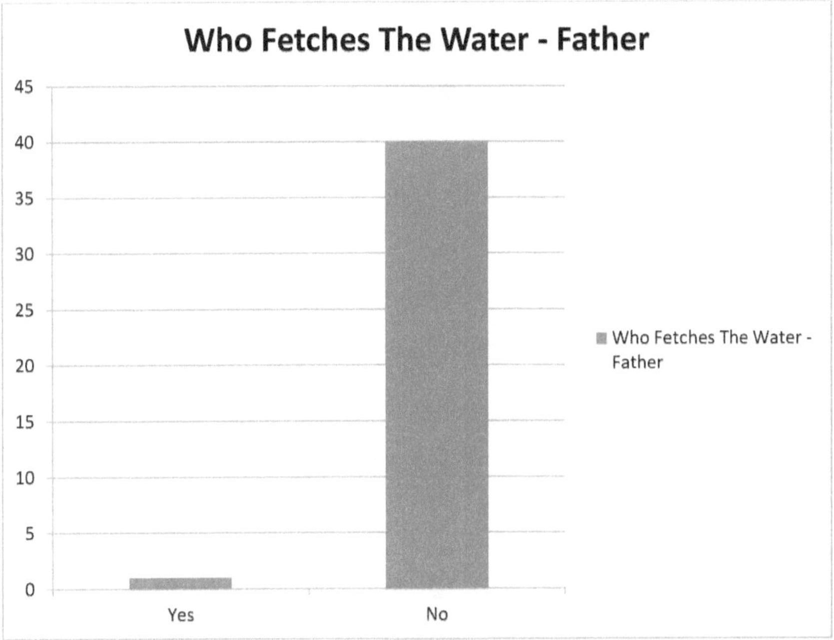

Note: graphic provided by the author

Part C of the fourth question asked about the respondents' views on female children bringing water to the household. 51.2 percent said it was true, while 48.8 percent said otherwise.

Figure 68. Who Fetches The Water—Female Child

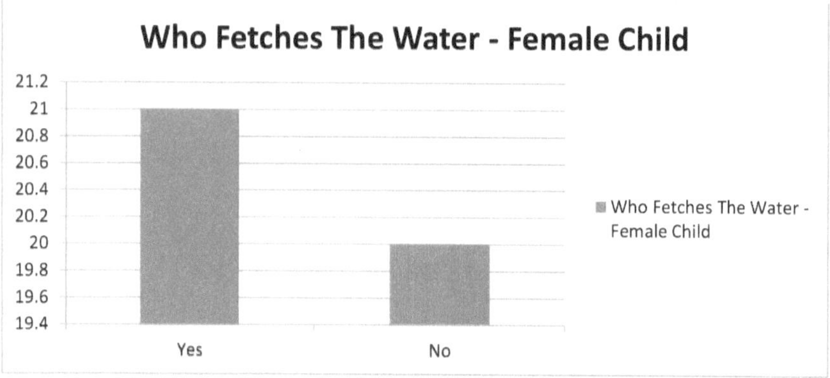

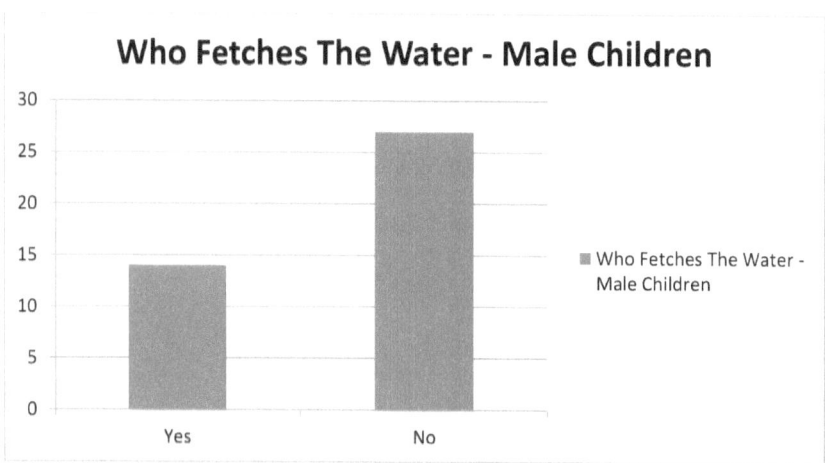

Note: graphic provided by the author

Part D asked the same question but regarding male children fetching the water for the household. 34.1 percent of respondents confirmed the male children brought the water, and 65.9 percent said otherwise.

Figure 69. Who Fetches The Water—Male Children

Note: graphic provided by the author

In the next two parts, the female and male house help was asked about bringing water to the household. For female house help, 12.2 percent responded affirmatively, while the remaining 87.8 percent responded negatively. For male house help, only 2.4 percent believed they should fetch the water, and 97.6 percent believed otherwise.

The fifth question asked the respondents whether they felt exhausted and overwhelmed with work during the day. 97.6 percent said they were overwhelmed, while only 2.4 percent said otherwise.

Figure 70. Overwhelmed with Work

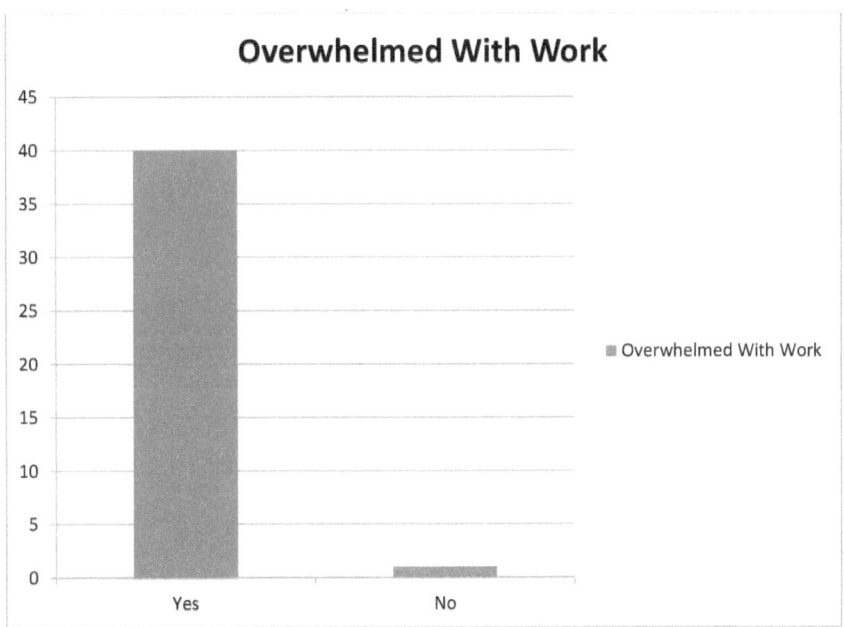

Note: graphic provided by the author

The next question asked about the time of the day when the respondents felt overwhelmed. 41.5 percent mentioned they were overwhelmed during morning hours (6:00–11:00 a.m.), 12.2 percent mentioned the afternoon hours (12:00–4:00 p.m.), and 43.9 percent said they were exhausted in evening hours (5:00–9:00 p.m.). The Y-axis represents the number of respondents, while the X-axis shows the time when the respondents feel overwhelmed.

Figure 71. Time of Day and Feeling Overwhelmed

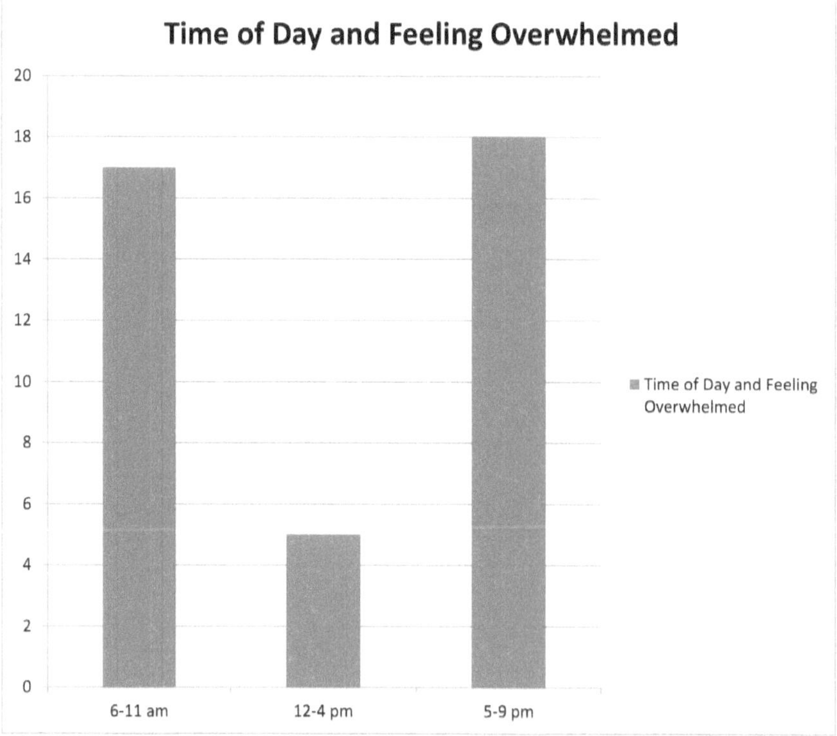

Note: graphic provided by the author

The next question regarded the quality and safety of the available water. 48.8 percent of the responses showed the water was safe for drinking, while 51.2 percent believed it was unsafe.

Figure 72. Is the Water Safe for Drinking?

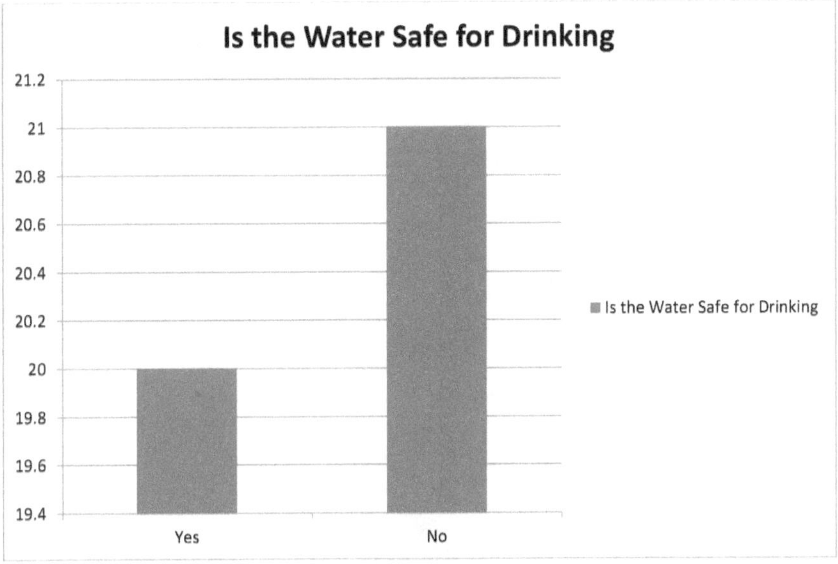

Note: graphic provided by the author

The following parts of this question explain the reasons behind the water being unsafe for drinking. Environmental cleanliness and the quality of the available water are absolutely vital for the population of South Sudan (Smith 2013). The first reason identified was the treatment of water. 22.2 percent did not respond to this part, 36.6 percent believed the water was unsafe because it was not treated, and 41.5 percent believed treatment was not a factor in the safety of drinking water.

Figure 73. Unsafe Drinking Water (Non-Treated)

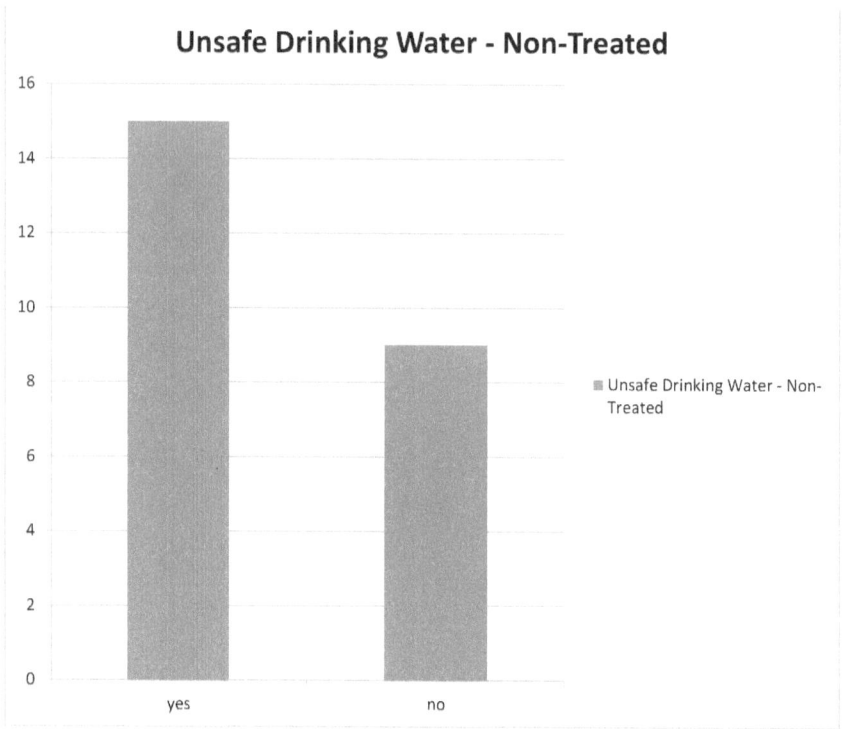

Note: graphic provided by the author

The next part mentioned water contamination as a factor of safety. 14.6 percent believed the water was contaminated, while 43.9 percent believed it was not contaminated.

Figure 74. Unsafe Drinking Water Contaminated

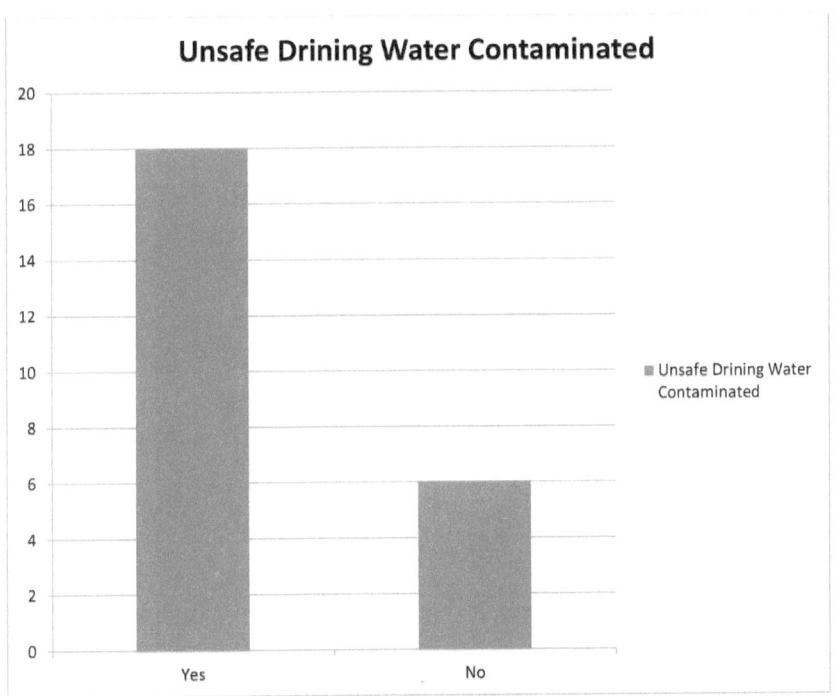

Note: graphic provided by the author

Then the respondents were asked about water safety regarding pollution. Only 2.4 percent believed the water was polluted, and 56.1 percent did not think so. Regarding unclean environment, 22 percent believed the environment was unclean, while 36.6 percent held an opposite view. As for the last option, broken water pipes and tanks, 26.8 percent believed this was the reason for unsafe water, and 31.7 percent did not agree.

Figure 75. Reasons for Unsafe Water

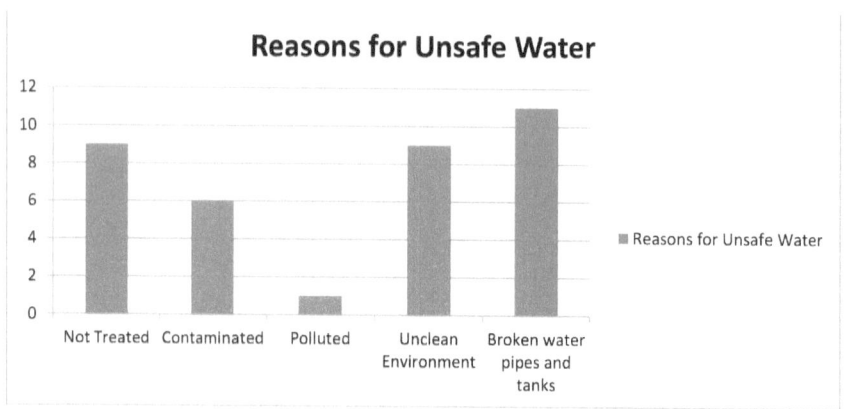

Note: graphic provided by the author

It can be seen in the above chart that most of the respondents believed the water was unsafe due to broken water pipes and tanks. The least probable cause of water being unhealthy seemed to be pollution. The eighth question asked for the respondents' feedback regarding the usage of fetched water. 90.2 percent responded it was used for domestic purposes, 4.9 percent said it was used in construction, and 2.4 percent claimed it was used for washing.

Figure 76. Usage of Fetched Water

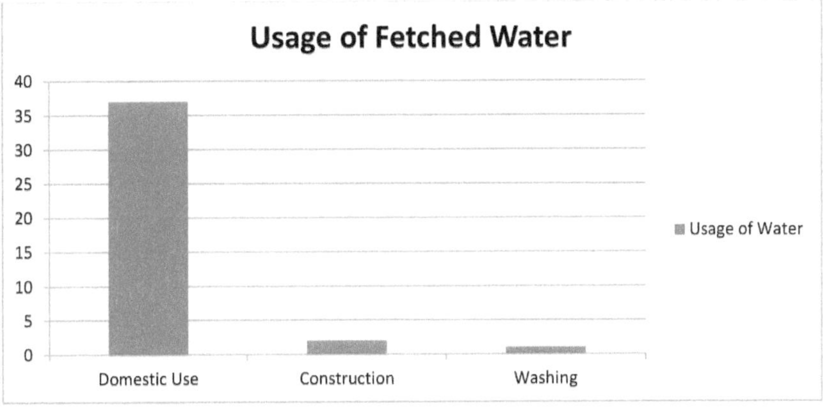

	Usage of Fetched Water	
Domestic Use	~37	
Construction	~2	
Washing	~1	

Note: graphic provided by the author

The next question asked the respondents about the importance of water kiosk projects in the communities of South Sudan. 92.7 percent considered it to be very important, while 7.3 percent believed it was important.

Figure 77. Importance of Water Kiosk Projects

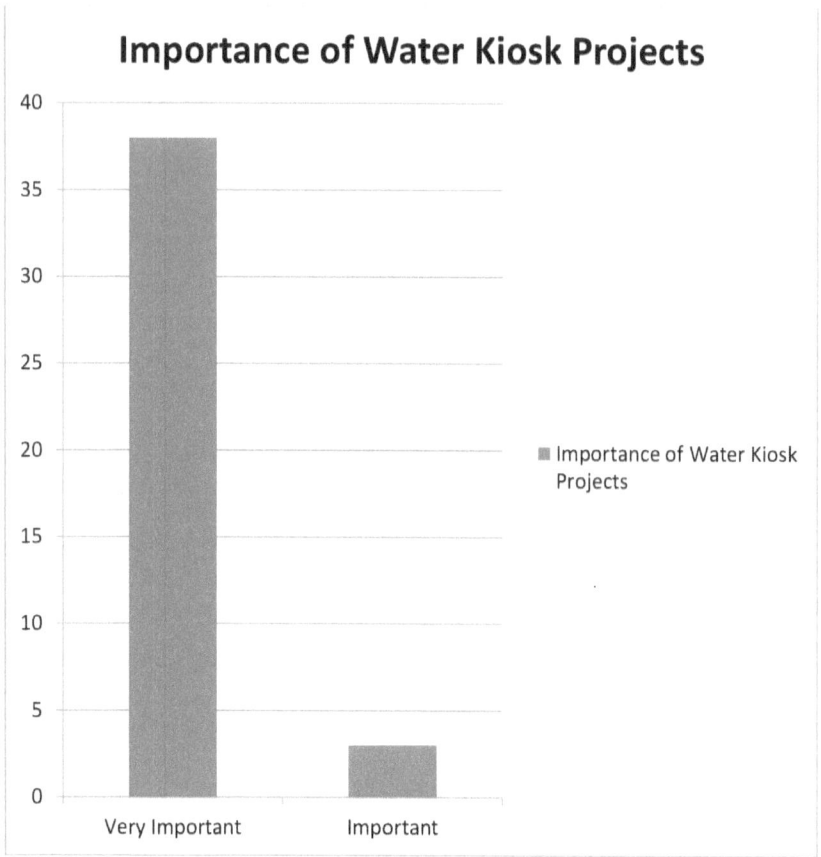

Note: graphic provided by the author

The next question was about the challenges of water kiosk projects. Several options were given to the respondents; the first option was lack of capacity identified as a challenge. 61 percent agreed it was a challenge, while 39 percent did not.

Figure 78. Lack of Capacity

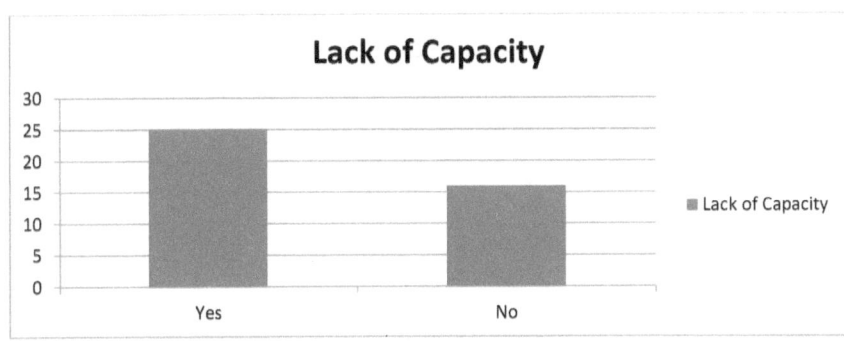

Note: graphic provided by the author

The next option was lack of regulations. 19.5 percent responded positively, and 80.5 percent responded negatively.

Figure 79. Lack of Regulations

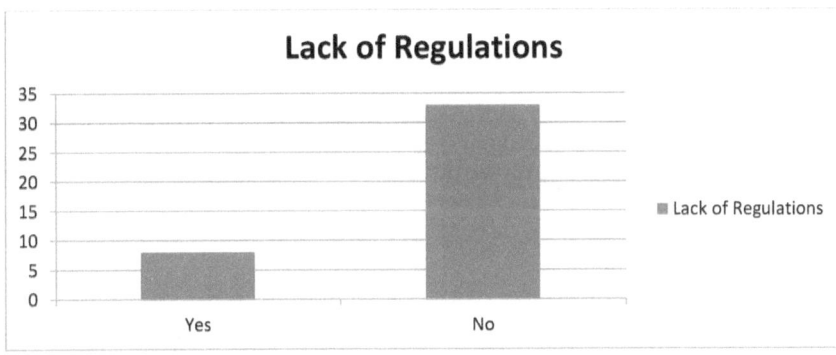

Note: graphic provided by the author

The next option was reluctance in paying bills. 22 percent believed this was a challenge, while 78 percent believed otherwise.

Figure 80. Reluctance in Payment of Bills

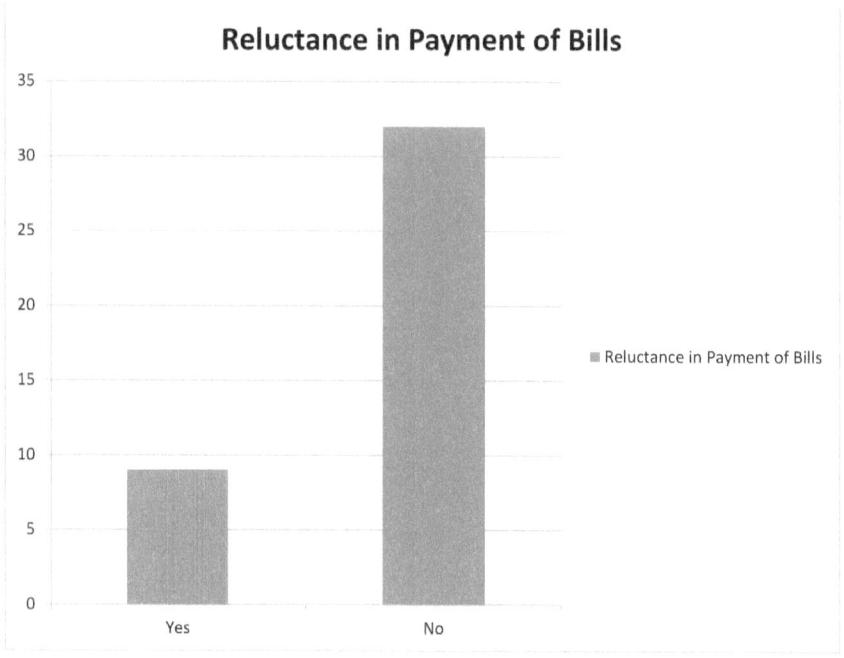

Note: graphic provided by the author

The next option was high demand as a challenge. 51.2 percent agreed it was a challenge, while 48.8 percent did not consider it a challenge.

Figure 81. High Demand

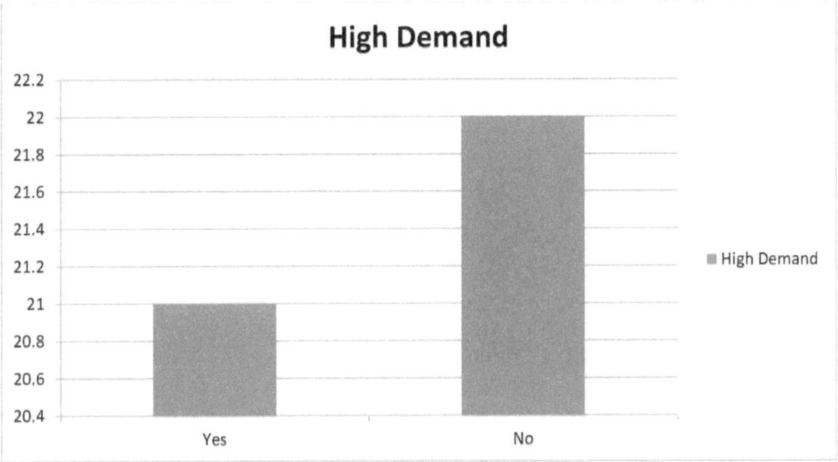

High Demand

(Bar chart showing: Yes ≈ 21, No ≈ 22)

Note: graphic provided by the author

Regarding unsafe water being a challenge, 9.8 percent agreed it was, while 90.2 percent disagreed. As for the option of poor management, 48.8 percent responded positively and 51.2 percent negatively. Finally, the last option offered was the theft of equipment as a challenge, and 14.6 percent considered it a challenge, while 85.4 percent thought otherwise.

The next question asked the respondents to choose potential solutions to these challenges. The first option given to them was capacity training, and 78 percent believed it could be a solution, while 19.5 percent disagreed. For the option of increase in water production, 75.6 percent responded positively, while 22 percent of responses were negative. The option of introduction of regulation and treatment of water had 22 percent of respondents who chose it, while 75.6 percent of respondents discarded it. As for redesigning the payment system,

9.8 percent believed it to be a solution, while 87.8 percent disagreed. Lastly, for the provision of security, 17.1 percent responded positively, while the remaining 80.5 percent responded negatively.

The last question asked the respondents whether risks threatened the water kiosk project. 97.6 percent of respondents believed it to be true; only 2.4 percent believed it was risk-free.

Figure 82. Water Project Under Risk

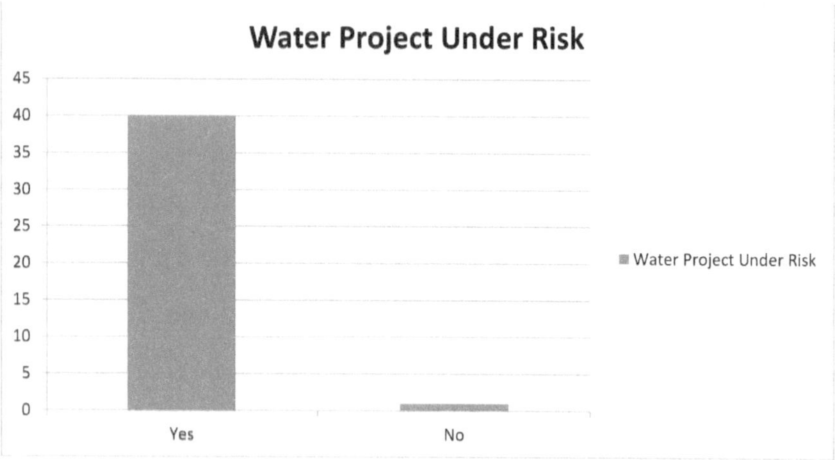

Note: graphic provided by the author

Different options were offered to the respondents. The first choice was risk associated with conflict of interest. In this regard, only 17.1 percent considered it to be a risk, while the rest, 82.9 percent, did not consider it a risk. The second option was the breakdown of equipment, and 90.2 percent considered it a risk, while only 9.8 percent responded negatively. The next option was cultural perception, and only 2.4 percent agreed it was a threat, while the rest, 97.6 percent, disagreed.

In the context of financial issues, 56.1 percent found it to be a threat, while 43.9 percent did not. 26.8 percent consider that administrative issues posed a threat to water projects, while 73.2 percent did not. The last option was the theft of equipment, and 31.7 percent considered it to be a risk, while 68.3 percent did not.

Summary of Findings from the Interviews

This section summarizes findings from semi-structured interviews, field observation, and document reviews on water kiosks in South Sudan. The interviews' purpose determined some of the underlining risk factors and challenges affecting water kiosks in South Sudan. Twenty interviews were conducted, amounting to 416.56 minutes, or 6.94 hours. The researcher also visited the site of the water projects and took pictures to aid the presentation of this report. The last part of this presentation summarizes the challenges.

Challenges of Water Projects

As a young country, South Sudan faces many challenges in the water sector. These challenges include but are not limited to infrastructural, economic, technological, legal, and political. Respondents attributed some of these challenges to power generators or solar used to pump the water to the reservoirs or tanks for distribution.

> The first big challenge is that the supply of the water from the main point is not regular. Sometimes during

> rainy season, when there is no sunshine for the solar to pump the water, there is no water. It takes us two to three days, sometimes even four days without water. This is one of the challenges we are facing. This water is not regular. (INTRV #17)

> The second challenge is about the payment of the fees. It is not done. Sometimes when you ask them (customers) to pay the money, they will say no; they do have money, but they are fetching water. So, it is a big problem. When I try to stop them, they will complain—where will they get water from? Like in our place here, there are only two water points. The main kiosks and this water point here in my house. This is one of the challenges we are facing in this village. (INTRV #17)

There were also complaints about the opening and closing hours of the water kiosk. Some customers stated the period was short, and sometimes customers did not get the water due to the flow.

> Sometimes customers do not get the water, especially if the flow of the water takes only a short period of time. Customer will go home without the water, and these also affect the management of water point;

sometimes, this leads to people scrambling for the water.

Reasons Why the Water Project Will Not Live Longer

Water is an essential commodity. The Sustainable Development Goal (SDG) 6 ensures the availability and sustainable management of water and sanitation for all. It states that by 2030, the world should achieve universal and equitable access to safe and affordable drinking water for all. To achieve this target, countries, governments, and developing partners and agencies need to work together to support and strengthen the participation of local communities in improving water and sanitation management. Some countries manage the water on the government level, while others have privatized water sectors for effective management.

In South Sudan, most community water projects are initiated in the form of projects specifically by NGOs with support from state and national government, using the concept of public-private partnership of build-and-handover to the community. At times, the build-and-handover concepts may pose a challenge to the facility's management. Respondents were asked whether the water projects would exist in the next few years to come. The following were some of the responses.

> …water will exist when this crisis stops; because it is not difficult to get the pumps. There are members of the community who contribute, even myself; I do

contribute a lot to make sure that the water project runs. (INTRV #1)

However, some respondents thought otherwise.

> The project will not exist after ten years because the solar panels which were brought have been stolen. There are no spare parts of the pumps. There were five boreholes dug but now only three (3) are functioning. The two stopped because they stole the solar panels. Also, one pump got burnt. The price of the pump is very expensive, and I doubt that the community can afford to buy one. (INTRV #16)

The Water Association or the Committee handles the management of the water. This group was formed to oversee the activities of the water projects. Such projects require qualified technical persons. However, one respondent had the following to say:

> The management is not capable to manage this water project. This is based on the capacity they have. They need training and set control of the water project. Since this is not there, I am not sure this water project can live for long. On the other hand, there are no professional technicians who can work on this water project when it is having a problem. This cannot make the water project stay for long. (INTRV #17)

The same respondent continued, "Apart from the water kiosks that were allocated by then, people have decided to connect it to their own houses like mine here. In the process, it will reduce the capacity of the water, and it will not live long."

Metering and Payment Methods

As Bonsor et al. (2015) illustrated, the concept of payment for water helps maintain the water project and remunerates employees managing the facilities. The approach taken focuses on decentralizing and commercializing the water supply and sanitation. This kind of arrangement is done in partnership with the community Water Association, a group tasked with the day-to-day water operations. The majority of members of such associations are employees from other institutions who support the cause of the community. Sometimes they do not have time to clearly focus on the community's programs. The use of meters and payment for water among communities used to free water have always met resistance and sometimes proved to be unsustainable. When the payment method is introduced, existing boreholes, streams, and rivers become alternative water sources for the communities. Many years of civil wars coupled with the world economic crises affected South Sudan's economy. Government employees can go five to eight months without receiving their salary. This situation has affected the economic status of the community members. Besides, jobless community members cannot afford to pay water fees when they live below poverty or on a dollar a day.

There some people, when they pay the monthly contribution, they want to bring their friends or neighbor to come and fetch water here. This increases the number of customers at each water point.

In terms of finance, when we don't have money to pay to the water management committee, it is another challenge to us because the customers are failing to pay their contribution. So, in this case, the water management will come and ask that they need the money. But at that time, we don't have money and, given the economic situation, it makes it hard for us to pay. This is another challenge.

Safety of Water for Drinking

According to Anke Peine Christian (2017), there is no plumbing work done in South Sudan. Because of that, many trucks are distributing drinking water in Juba and other states or counties in the country. Infrastructure development is poor because of the decade of civil wars. International organizations and partners such as GIZ have come to support physical infrastructure development to provide clean drinking water to the communities. Local institutions must be established and trained to help provide water to the communities.

Respondents across the community are divided on the safety of the drinking water. Some respondents neither view the water as safe and treated, while others feel the water is not safe or treated. Those who see the water as unsafe for drinking attributed it to the rise of waterborne diseases such as diarrhea, worms, and typhoid in the state hospitals.

> This water is not safe because you can see with your naked eyes how dirty it is. Also, when it is examined in the lab, you will find there are bilharzia and other worms. Children always play around the water points, making it dirty. The environment is very dirty. It is not clean. The pipes are not standard because they are not fixed well. (INTRV # 17)

Another respondent said, "Before, water was being treated, but nowadays it is not treated. Again, most of the pipes are breaking, and some dirt gets into the pipes, and when the community drinks the water, it can harm them. So that is why I am saying the water is not safe" (INTRV # 8).

When the same question was asked to another respondent, this was the response,

> Yes, the water is very good and clean water. There is no difference between this water and the mineral water being sold to people in the market. People drink this water for some time, and then the Water Association will inform the community that they are putting medicines into the water. There will not be

water for a day or so. During that time, they clean the tank and treat the water. So, it is very clean water. They always clean the tanks and treat the water. (INTRV # 14)

Another respondent stated, "It is safe and tasty, you can have it." The respondent said the community knows it, but ignorance is a problem (INTRV # 19).

Security of the Water Projects

According to the ECHO fact sheet (2021), increased insecurity in South Sudan obstructs charitable services in the country. Other agencies see the situation as volatile. This situation further affects the water sectors, especially in the rural areas. The equipment is often destroyed, especially during intercommunal fighting. During the interviews, the majority of the respondents reported that an unknown gunman often steal some water kiosk equipment.

> The security of the equipment is very poor. Recently we have seen that people go and remove solar panels; they steal them. Once solar panel is removed from one part, then the capacity of the power is reduced, and this can make capacity of the water to stop at any time because there is no proper security. I have learned that there are some boreholes which do not supply water because the wells are drying up. (INTRV # 17)

> Theft of solar will affect the existence of the water. We need to take care of it and ensure that it used in the right way. This will help the water to stay long otherwise this water is at very high risk of collapsing. (INTRV # 17)

Another respondent said,

> There are some water pumps which are being protected, but others are not. We also wonder why because it was in the plan that security should be provided to protect the water pumps. Maybe the money given to the security guard is not enough, we do not know. We even do not know where the need for the people money collected goes. That money should be paid to the security of the water. (INTRV # 11)

Risks of Managing Community Water Kiosks

Risk management is a process. It helps in identifying, assessing, controlling, and reviewing controls or measures to mitigate the effects. On the other hand, risk is defined as an unforeseen problem that affects the implementation of any project. Effective risk management suggests reducing both the possibility of a risk happening and its potential impact on the community water kiosks. In the case of the water sector, insufficient water could pose an immense risk to the community, whose basic needs would have been curtailed.

Irrespective of the challenges facing community water, the water demand is still very high. The management is often overwhelmed by the work and sometimes could not pay attention to water breakdown in residential areas. Some respondents stated that many pipes break down, posing high risks of waterborne diseases. One respondent, a medical doctor, said,

> The management does not do good work. Some pipes were spoilt in the hospital, but there is no one to come and repair them. We have been running up and down to look for them, but we could not get them, the water was running out, and no one to take care of it. This clearly shows that the capacity of those managing the water needs to be improved. (INTRV # 6)

Conclusion and Recommendations

Introduction

South Sudan got its independence in 2011 after going through a war for independence against the North. Soon after its independence, it saw another civil war against the rebels with different conflicts. Before the independence, the developmental administration was handled by the North, which left the South with multiple problems to handle without a developed system. Because of its infancy and the two wars, South Sudan faced a severe economic crisis worsened by the refugees' return to their homeland once peace was established.

Although a newly independent country faces various issues, the water supply and management is the greatest one. Water is the basic necessity for everyone, and its availability, along with its quality, are of an essence. 15 percent of the water requirements are fulfilled by the White Nile River, which flows through South Sudan, and almost 27 million cubic meters of water flow through the country through this river. However, only 40 percent of the population has access to water for basic needs. Much water is lost due to evapotranspiration, a process whereby evaporation comes from the soil and transpiration from the leaves of the plants. The White Nile is not only accommodating South Sudan. Uganda, Tanzania, Kenya, Rwanda, and Congo also consume its water. After South Sudan, the river flows into Egypt and Sudan, which also have their claim on its water. The water dispute among the African countries has been a historical battle.

There is an utter need to introduce new water kiosk projects and aquifers in South Sudan. The government, community, and international organizations are making efforts to solve the water issue in the country. The MWRI is one of the leading government bodies responsible for the water supply and management in the country. The main purpose of MWRI is to develop policies, improve the quality of water, and monitor the installation of new projects at the local as well as central level. Another organization, SSUWC, works under the Ministry and also handles the water supply and management in urban areas. When it comes to private and community-based projects, the Ministry of Housing and Planning is also involved. South Sudan is adopting two major water policies. First is the South Sudan Water Policy, formulated in 2007 and adopted in 2009. The central principle of this policy is that every citizen has a basic right to good-quality water and sanitation. This policy also encourages private water suppliers and community developers. This policy also recommended that different water organizations handle the urban and rural water supply. In 2011, another policy document was introduced, the Water, Hygiene and Sanitation Strategic Framework for South Sudan.

Many international organizations are actively involved in the water project development and improvement of water supply in the country. These organizations include USAID, World Bank, and AfDB. Two different projects of USAID, Global Water Development Strategy and Transition Plan for South Sudan, handles South Sudan water. According to USAID, various areas need to be modified to solve the water issue in South Sudan. Similarly, the AfDB has also pointed out the areas where improvements are required for enhancing the availability and quality of water. Entrepreneurial innovations should be used in a newly independent country facing financial constraints. It is vital to increase water supply organizations' capacity along with water supply and sanitation improvement. Building business relationships with private and community developers is also required. It is important the new water projects are designed and implemented so they are sustainable. The already-installed water projects need upgrades and maintenance. There is an urgent requirement for providing training and awareness to the local community regarding water use and payment of bills.

After conducting this research, it is concluded that the risks and challenges faced previously by water kiosk projects are still present. It was also observed that the approaches to handle and mitigate these risks are still almost the same. Unless innovative ideas are introduced, it will be difficult to solve the water issue in South Sudan. Using modern business tools and techniques, including cost management, operational management, risk management, and project management, can greatly enhance the performance of the water kiosk projects. Lean Six Sigma approach is one of the best ways to remove waste and flaws in the country's water operations.

It has also been seen there are different kinds of challenges that water projects face, and these are political, financial, security, and technical issues. The communication gap between the local and foreign individuals attached to the water projects is also another issue causing hindrances to the success of water projects. Training and education are required not only for bridging this communication gap but also for creating awareness among the local community with respect to water usage. Even government bodies and international organizations are having trouble communicating and synchronizing with each other. Various institutes are handling the same task, causing a waste of resources. Most of the employees and water institutes do not have a proper job description, which causes delays in operations. There has to be a proper system that monitors the maintenance of the existing water kiosk projects and supports new ones. Sustainable water kiosk projects are required in the country by taking appropriate actions and implementing innovations.

Discussion of the Findings

Water is a natural resource essential for the existence of human beings and the survival of other plants and animals. Water sustains life and ecological systems and is a vital resource for social, cultural, and economic development and the sustainability of any community (Omar et al. 2017). However, developing countries face significant challenges in sustaining water quality and ensuring a sufficient freshwater supply to meet the growing demand. Similarly, access to clean water has remained a significant challenge for the majority of the African population. Therefore, there is a need for community water projects, such as water kiosks, to ensure safe drinking water.

Countries such as South Sudan are among the countries in sub-Saharan African facing a growing shortage of clean and consumable water.

There is universal consensus among researchers that community water projects are crucial and play a significant role in sustaining life in the community. Community water projects, such as kiosks, ensure an adequate supply of clean and quality water for consumption. They act like storage centers where water is temporarily stored, cleaned, and purified for the entire community's benefit. Nyakwaka and Benard (2019) agree that community water projects require constant monitoring and proper management to reduce the risk of community water projects. Water kiosks have become vital resources in different communities and can satisfy community water needs if implemented and properly managed. Community water projects, such as water kiosks, enable the residents of various regions to meet their basic needs. Since water is considered one of the essential basic needs and vital human rights, water stalls/kiosks provide access to clean water and enhance the quality and standard of life (Stoyanova et al. 2018).

Similarities and Differences with Other Studies

Researchers agree numerous risks are associated with community water projects that threaten their existence and prevent them from achieving their objectives in society. Despite the numerous benefits associated with community water projects, there is a lack of proper management of water kiosks (Omar et al. 2017). This has significantly contributed to the collapse of numerous water projects in developing countries, such as South Sudan. Proper management, coupled with effective governance of community water kiosks, is essential to ensure water safety and efficiency. Stoyanova et al. (2018) emphasize this can also contribute to a massive reduction of environmental destruction and stop the spread of preventable diseases. One way to ensure that public water projects remain beneficial to the community is to establish strong and effective policies to manage risks.

Studies have identified various risks involved in the management and supervision of public water projects in developing countries. Although researchers have identified similar risks, they have also concurred that some risks are natural, making them difficult to control or prevent (Stoyanova et al. 2018). However, most researchers have concurred that various risks associated with public water projects can occur due to human factors, including human movement and actions (Omar et al. 2017). Several public water projects have collapsed in Africa because different community water management has failed to involve the participation of local communities. As a result, the project management procedures and processes fail to consider the community needs leading to the collapse of the water projects.

Other research findings have unanimously agreed that public water projects face significant risks. They agree that risk exists, but it is important to determine the nature of risk to develop a specific remedy for each risk. The findings of Stoyanova et al. (2018) have categorized the risk associated with public water projects as legal risks, political risks, technical risks, management risks, security risks, and financial risks. This is similar to the findings of this study, which has also associated the same risks to the community water project. Other researchers have also identified these risks as potential risks contributing to the collapse of public water projects in rural areas in developing countries. Legally, there are established regulations that the projects must comply with before and after the commencement of the operations. Nyakwaka and Benard (2019) concur that failure to comply with legal requirements poses a potential threat to the successful implementation and management of public water projects.

Financial risks are also another key factor influencing the management of community water projects, according to the study findings. Proper management and supervision of the community water kiosks require adequate financial resources, without which the project faces the risk of collapsing. Management risk involves a lack of adequate management skills to oversee the operations of public water projects. The findings of this study and that of Stoyanova et al. (2018) agree that poorly managed pubic water projects have collapsed or are on the verge of collapsing. Findings by different researchers have agreed that community water projects also face massive security risks. There is the risk that the projects can be vandalized by unauthorized personnel, leading to the destruction and collapse of the projects (Nyakwaka and Benard 2019).

However, some researchers have deviated from the findings presented in this dissertation project. While the project finds no significant association between political factors and the collapse of public water projects, other researchers, such as Harvey and Reed (2006), have linked political risk to the collapse of public water projects. Researchers with different opinions conclude there are also political risks associated with implementing, managing, and supervising public water projects. South Sudan and other developing regions are usually characterized by unending political issues, including the struggle for government control and key resource centers. Therefore, politics pose a potential threat to the proper management of public water projects.

Importance and Implications of the Findings

While other studies have focused on the sustainability of public water projects, this research has provided a different perspective by presenting findings that focus on the risks associated with the collapse of these projects. Therefore, the findings of this study are most important than other studies because it has highlighted the risk factors contributing to the constant failure of community water projects. The study has focused on analyzing the risks involved in managing water kiosks in rural areas of South Sudan. It focused on the reasons behind the massive failure of several community water kiosks, which makes it more important than other studies. It has also provided recommendations about the ways through which water vendors can control the risk. Therefore, this study has focused on areas that other researchers have not fully explored. This implies that findings are more important and can significantly contribute to the management of water kiosks than other research findings.

The study findings will be used by the various governments, community groups, and business organizations, including regional and international public, to ensure the risk associated with water kiosks is prevented. It will establish strategies for monitoring and reporting the community water project issues in developing countries to develop effective measures to address them. The study's findings will also be used by the South Sudan government and other governments, including the employing agencies, to reinforce their public relation initiative to address the challenges associated with water projects to make them more effective and productive. Similarly, the federal government can use the study's findings alongside other initiatives to determine the best way to communicate with the community while conducting outreach programs and campaigns. The findings can also be useful while implementing community-based infrastructure projects in conjunction with other players to develop effective policies. Other researchers can also use this study to identify the gaps that require future analysis.

Researcher's Opinion on the Findings

Personally, the findings of this study were accurate, credible, and reliable. This is because the study has considered contributions from various scholars with significant knowledge about the public water project. The study findings were arrived at after extensive investigation of various stakeholders with a direct link to the community water project. I agree with the finding since it points out the true cause of problems in numerous water projects in developing countries. The findings are practicable, meaning they can be applied in real project management situations. Several local and international organizations are directly engaged in water projects; therefore, I believe these findings will play a key role in facilitating their efforts to manage the projects. I believe the findings can be applied in any country, irrespective of its economic strength or size.

Contribution to Knowledge

This research was conducted so all the previous projects and literature were thoroughly studied to find solutions for the risks involved with the water projects. All this data is collected, analyzed, and presented in this study, enabling future researchers to gain all the knowledge regarding the South Sudan water issue from this single document. It highlighted various new areas not shown in previous studies, like identifying the lack of risk, cost, and operations management in the system. It also highlighted the important fact that the communication gap between the local and foreign stakeholders is causing a significant hindrance in the process. It is also recommended, for the first time in this study, that only one company should be fully authorized to control the water management, opposing the South Sudan water policy in this particular point.

Contribution to Practice

Public water projects are present in every country around the world, especially in developing countries. The major problem facing the success of public water projects includes the mismanagement of administrative duties. This has massively contributed to numerous risks, leading to the collapse of water kiosks in developing countries. The study will be used to identify the risk involved in water safety resulting from the mismanagement of administrative duties. The recommendations provided will be applied to ensure the risks identified are adequately addressed. Therefore, this study will be used to control, operate, and manage various public water projects in developing countries.

Limitations of the Research

During this research, a lot of qualitative and quantitative data were collected, but the firsthand observation could not be completed as the researcher could not visit all the sites on the ground. To mitigate the effect, the firsthand observation of the interviewees was collected and included in the progress of this research. The official documents of the NGOs and other community organizations were inadequate.

As South Sudan recently gained its independence, a small amount of data was available regarding water issues in the country.

Research Objectives

The research aimed to determine the risks involved with water safety due to the mismanagement of administrative duties. It also determined the risks the community faces due to the lack of governance and an outdated approach. This research's objective was to determine the methodological errors and how community projects can face dangers. After gathering empirical data and literature from various sources, the study proposed a model that could be used to manage water projects: the use of modern business tools and techniques, the safe water enterprise, and the Lean Six Sigma approach to improve water project management in South Sudan.

Future Research

After applying the action plan presented in this research, it is important to keep getting feedback, and the cycles of action research are continued until the desired results are obtained. There is also the part of moving the water from the evapotranspiration region to safe storage where it is not lost. The addition of that water to the system can significantly enhance the water supply of South Sudan. Research on upgrading the country's irrigation system is also how water issues can be further addressed.

Recommendations

Researching risk management in water projects in South Sudan, where there is no evidence of risk management tools and policies, is a prodigious encounter. Projects have time limits, and no concrete procedures are put in place. Many projects are bound to collapse, more so the projects that do not have entrepreneurial approaches. The study findings revealed several factors affecting water projects.

- ❖ **Allocation of Water Kiosks and Site Identification Process:** Although site identification depends largely on geological surveys conducted to detect underground water levels, it was reported not inclusive. Involving the communities in the process would minimize some of the risks identified.
- ❖ **Equipment and Installation:** Arguably, community members were not part of the installation. Therefore, it is recommended that community members should be involved as they are the benefices who will be in charge of the projects in the years to come. Equipment such as pipes, pumps, and tanks should be of good quality and installed in a protected manner to avoid breakdowns all the time. Instead of smaller pipes, larger pipes should be used to supply water.
- ❖ **Finances, Budget, and System of Water Fees Collection:** Lack of budget and proper procedures to collect water fees was also another risk factor reported very high. It is therefore recommended that an appropriate means of financial allocation

and management be established. Financial constraints must be removed with the help of cost management and entrepreneurial projects. This would help pay security staff and reduce the threats related to the security of water kiosks. Payment systems need to be redesigned and made easier for the community.

- ❖ **Communication and Cultural Perceptions:** Community awareness regarding water usage should be promoted, and entrepreneurial concepts should be explained to the community. Communication gap between the local and foreign stakeholders must be enhanced for mutual understanding and cooperation. Cultural perception regarding water usage needs to be changed through education and awareness.

- ❖ **Regular Maintenance and Spare Parts**: Regular maintenance and provision of spare parts should be made available, as it has been a big challenge in the operations of the water kiosks. This would prolong the life span of the water projects and avoid constant breakdown.

- ❖ **Need for Good Governance:** Good governance is a vital component and critical for project implementation. Management and administration of the water kiosks need a lot of improvement and modification. There is also a great need for urban planning, especially for future infrastructure of the country. This should be done so the current installations are connected and smaller lengths of pipes are required to connect the new dwellings. Risk management theories should be applied to tackle political, financial, technical, management, and security risks associated with water kiosk projects. A risk mitigation plan should be formulated before starting a new water kiosk project.

- ❖ Functions of the various arms of the Government: Although the roles and functions of the national government, state, and county government are clearly defined, there was interference in the functions perhaps because of the financial attachments. Thus, the different arms of the governments in charge of water sector should be engaged in the water projects, and the political leaders consulted to avoid hindrances and interferences in the water projects. Regular meetings with various stakeholders, especially the community and political authorities, are paramount.

- ❖ Concise and clear water policies: the report revealed that in some counties or states, several organizations were allocated water projects. This situation created conflicts among some organizations resultant to substandard services. Therefore, multiple organizations should not be allocated to a single task or area of operation. In the event where more than two organization are implementing water project in the same area, there duties and responsibilities should be clearly stipulated. Water policies should be strictly followed, visible in the water projects, and in compliance with the community guidelines. This would eliminate competition between private water vendors and suppliers.

- ❖ Hygiene promotion and education: Some community water projects were very unhealthy. Some projects had untreated water because water guard and treatment was not performed for many months. In the design of the water projects, water treatment plans should be integrated with the water supply projects. Chlorine should also be administered regularly.

❖ Capacity development of water staffs: Training of the managers and workers should be done systematically. If physical training centers are not possible due to financial constraints, virtual learning should be used. Also, the number and capacity of water kiosk projects should be increased to solve the water crises in the Republic of South Sudan.

❖ **Further Recommendations:**
- Additional funds must be generated by introducing innovative ideas for the cost and time reduction of the processes. The bill collection method should be improved, and bills should be subsidized so they are within the purchasing power of the consumer.
- Project management and operations management techniques should be implemented on water projects to increase their efficiency. During the installation and operating hours of the water kiosks, professional supervision is essential. In addition, the installation and maintenance of solar systems should be encouraged for longer sustainability of water projects.
- Community meetings should be held to spread awareness and collect feedback, and before initiating a water project, consensus should be made between the community dwellers,

and precautions must be taken during the excavation of the water projects.

- Water quality should also be improved by implementing innovative measures and technology. Clean and strong pipes should be used to supply the water. The storage tanks should also be cleaned and maintained regularly. The water supply should be tested for contamination at regular intervals.

- The breakage of pipes and tanks should be regularly monitored because it is the biggest threat to water cleanliness. The environment around the water projects should be kept clean at all times.

- Before installing new water projects, it must be made clear that there is no conflict of interest. And after implementing actions, feedback should be collected, and the plan should be modified according to feedback.

- Boreholes and open wells should also be promoted to the rural population, and hand pumps should be installed near small villages instead of installing water supply lines.

- The number of plantations must be increased in the country to make rains more probable. Canals and ponds should be encouraged to transport the water away from the region of evapotranspiration, which causes a lot of water to go to waste. And underground storage tanks in households should be encouraged.

- The country's irrigation system should be improved to solve the water availability problem.
- Theft of equipment must be stopped by increasing security and using guard dogs.

Summary

This chapter has provided a review and summary of the research findings and the extent to which the objectives were achieved. A conclusion has also been presented, limitations have been acknowledged, and further research areas were proposed. Various challenges are facing the water sector in South Sudan, with some relating to the current political, economic, and humanitarian situation of the country. The instability of the government, the massive level of corruption, and lack of basic social infrastructures have further acerbated the provision of basic social services to the citizens.

In summary, the research has proposed key business tools and approaches that could aid the water problems in South Sudan. These approaches include using modern business tools and techniques, including cost management, operational management, risk management, and project management, to greatly enhance the performance of water kiosk projects. It is also proposed that the Lean Six Sigma approach is one of the best ways to remove waste and flaws from the country's water operations.

List of Abbreviations

AfDB	African Development Bank
CAFORD	Catholic Agency for Overseas Development
CAR	Central African Republic
CBD	Community-Based Development
CBM	Community-Based Water Management
CD	Capacity Development
CDD	Community-Driven Development
COVID-19	Coronavirus
DRC	Democratic Republic of Congo
ECHO	European Commission Humanitarian Aid
GDP	Growth Domestic Products
GIZ	German Development Cooperation
LCA	Life Cycle Analysis
MDG	Millennium Development Goals
MWRI	Ministry of Water Resources and Irrigation
NGO	Nongovernmental Organization
SDG	Sustainable Development Goals
SSNBS	South Sudan National Bureau of Statistics
SSUWC	South Sudan Urban Water Corporation
SWE	Safe Water Enterprise
TWWMP	Tambura West Water Management Project
UKAID	United Kingdom Agency for International Development
UNDP	United Nation Development Programme
UNICEF	United Nations International Children Education Fund
UNOPS	United Nations Office of Project Service
USAID	United States Agency for International Development
WASH	Water Sanitation and Hygiene
WMA	Water Management Association

WMC	Water Management Committee
WVI	World Vision International
WOYE	Women and Youth Empowerment
YUMASCO	Yambio Urban Water Management and Sanitation Company

References

African Development Bank. 2012. "South Sudan: An Action Plan." Accessed December 20, 2019. https://www.afdb.org.

Akosa, G., R. Franceys, P. Barker, and T. Weyman-Jones. 1995. "The efficiency of water supply and sanitation projects in Ghana." *Journal of Infrastructure Systems* 1(1): 56–65.

Alaerts, G. J., and J. Kaspersma. 2009. "Progress and challenges in knowledge and capacity development." *Capacity Development for Improved Water Management*, 3.

Allan, J. A., ed. 2012. *Handbook of Land and Water Grabs in Africa: Foreign Direct Investment and Food and Water Security.* Routledge.

Anke Peine Christian. 2017. "Long-term structural development." https://www.dandc.eu/en/article/despite-civil-war-south-sudan-GIZ-working-infrastructure-improvement-supply-water-people.

Awadh, A., and A. Saad. 2013. "International Review of Management and Business Research." *Impact of Organizational Culture on Employee Performance* 2(1): 168–175.

Barrett, M., A. C. Farr, T. Wilson, and D. Crane. 2015. "Assessing graduate capability development through University STEM community engagement programs."

Baumann, E., P. Ball, and A. Beyene. 2005. "Rationalization of Drilling Operations in Tanzania." *Review of the Borehole Drilling Sector in Tanzania.*

Bessant, J., and J. Tidd. 2016. *Innovation and Entrepreneurship,* 3rd ed. London: John Wiley and Sons.

Bey, V., P. Magara, and J. Abisa. 2014. "Assessment of the Performance of the Service Deliver Model for Point Water Sources in Uganda." Final Research Report, WASH.

Bonsor, H. C., N. Oates, P. J. Chilton, R. C. Carter, V. Casey, A.M. MacDonald, and M. Bennie. 2015. "A Hidden Crisis: Strengthening the Evidence Base on the Sustainability of Rural Groundwater Supplies: Results from a Pilot Study in Uganda."

Budds, J., and G. McGranahan. 2003. "Are the debates on water privatization missing the point Experiences from Africa, Asia, and Latin America." *Environment and Urbanization* 15(2): 87–114.

Burns, P. 2013. *Corporate Entrepreneurship: Innovation and Strategy in Large Organisations,* 3rd ed. Palgrave Macmillan.

Carr, G., G. Blöschl, and D. P. Loucks. 2012. "Evaluating participation in water resource management: A review." *Water Resources Research* 48(11).

Carter, R. C., and I. Ross. 2016. "Beyond the 'functionality' of handpump-supplied rural water services in developing countries." *Waterlines* 35(1): 94–110.

Carter, R. C., S. F. Tyrrel, and P. Howsam. 1999. "The impact and sustainability of community water supply and sanitation programs in developing countries." *Water and Environment Journal* 13(4): 292–296.

Cohen, L., L. Manion, and K. Morrison. 2017. *Research Methods in Education*, 8th ed. London: Taylor and Francis.

Coldwell D., and F. J. Herbst. 2004. *Business Research*. New York: Juta Academic.

Collins, H. 2010. *Creative research: the theory and practice of research for the creative industries*. London: Thames and Hudson.

Conway, D., E. Allison, R. Felstead, and M. Goulden. 2005. "Rainfall variability in East Africa: implications for natural resources management and livelihoods." *Philosophical Transactions of the Royal Society A: Mathematical, Physical and Engineering Sciences* 363(1826): 49–54.

Creswell, J. W., and D. L. Miller. 2000. "Determining Validity in Qualitative Inquiry." *Theory into Practice* 39(3): 124–131.

Dagdeviren, H., and S. A. Robertson. 2011. "Access to Water in the Slums of Sub-Saharan Africa." *Development Policy Review* 29(4): 485–505.

Davis, R., R. Campbell, Z. Hildon, L. Hobbs, and S. Michie. 2015. "Theories of behavior and behaviour change across the social and behavioral sciences: a scoping review." *Health psychology review* 9(3): 323–344.

De Cecco, P. 2012. "Rural Water Development in Sub-Saharan Africa: A Comparative Study between Uganda and Tanzania."

Dell'Angelo, J., P. F. McCord, D. Gower, S. Carpenter, K. K. Caylor, and T. P. Evans. 2016. "Community water governance on Mount

Kenya: an assessment based on Ostrom's design principles of natural resource management." *Mountain Research and Development* 36(1): 102–116.

Dodgson, M., and G. Gann. 2010. *Innovation: A Very Short Introduction.* Oxford University Press.

Edwards, M. 1997. "Organizational learning in non-governmental organizations: What have we learned?" *Public Administration and Development: The International Journal of Management Research and Practice* 17(2): 235–250.

Effah Ameyaw, E., and A. P. Chan. 2013. "Identifying public-private partnership (PPP) risks in managing water supply projects in Ghana." *Journal of Facilities Management* 11(2): 152–182.

Eguavoen, I. 2008. "Changing household water rights in rural northern Ghana." *Development* 51(1): 126–129.

Emmett, T. 2000. "Beyond community participation? Alternative routes to civil engagement and development in South Africa." *Development Southern Africa* 17(4): 501–518.

Etongo, D., G. Fagan, C. Kabonesa, and B. Asaba. 2018. "Community-Managed Water Supply Systems in Rural Uganda: The Role of Participation and Capacity Development." *Water* 10(9): 1271.

Fernando, N., and W. Garvey. 2013. "Republic of South Sudan: The Rapid Water Sector Needs Assessment and a Way Forward."

Fielmua, N. 2011. "The role of the community ownership and management strategy towards sustainable access to water in

Ghana (A case of Nadowli District)." *Journal of Sustainable Development* 4(3): 174.

Financial Management of Water Supply and Sanitation. "1 Water supply- Economics 2 Sanitation, Economics 3 Cost control."

Foster, T., S. Furey, B. Banks, and J. Willetts. 2019. "The functionality of handpump water supplies: a review of data from sub-Saharan Africa and the Asia-Pacific region." *International Journal of Water Resources Development*: 1–15.

Franklin, C., and M. Ballan. 2001. "Reliability and validity in qualitative research." *The handbook of social work research methods* 4: 273–292.

Garn, H. A. 1997. "Lessons from large-scale rural water and sanitation projects: Transition and innovation." http://www.up.org/pdfs/working_projlesson.pdf.

Gasson, C. 2017. "A New Model for Water Access: A Global Blueprint for Innovation." Retrieved December 18, 2019. http://www.globalwaterleaders.org/water_leaders.pdf.

Gibb, A. 2000. "Corporate Restructuring and Entrepreneurship: What Can Large Organisations Learn from Small." *Enterprise and Innovation Management Studies* 1(1): 19–35.

Giné, R., and A. Pérez-Foguet. 2008. "Sustainability assessment of national rural water supply program in Tanzania." *Natural Resources Forum* 32(4): 327–342.

Golafshani, N. 2003. "Understanding reliability and validity in qualitative research." *The qualitative report* 8(4): 597–606.

Goldman, M. 2007. "How 'Water for All!' policy became hegemonic: The power of the World Bank and its transnational policy networks." *Geoforum* 38(5): 786–800.

Gopalakrishnan, N. 2010. *Simplified Lean Manufacture: Elements, Rules, Tools and Implementation*. New Delhi: PHI learning private limited.

Guide, A. 2001. *Project management body of knowledge (pmbok® guide)*. Project Management Institute.

Gumbo, B. 2004. "The status of water demand management in selected cities of southern Africa." *Physics and Chemistry of the Earth*, Parts A/B/C, 29(15–18): 1225–1231.

Harvey, P. 2008. "Poverty Reduction Strategies: opportunities and threats for sustainable rural water services in sub-Saharan Africa." *Progress in Development Studies* 8(1): 115–128.

Harvey, P. A., and R. A. Reed. 2006. "Community-managed water supplies in Africa: sustainable or dispensable?" *Community Development Journal* 42(3): 365–378.

———. 2003. "Sustainable rural water supply in Africa: Rhetoric and reality."

Haysom, 2006. "A Study of the Factors Affecting Sustainability of Rural Water Supplies in Tanzania."

Hemson, D. 2002. "'Women are weak when they are amongst men': women's participation in rural water committees in South Africa." *Agenda* 17(52): 24–32.

Hillson, D. 2017. *Managing risk in projects*. Routledge.

Holloway, I., and S. Wheeler. 2002. *Qualitative research in Nursing.* Blackwell Science, Oxford.

Hopkinson, M. 2017. *The project risk maturity model: Measuring and improving risk management capability.* Routledge.

Islam, S., and L. Susskind. 2015. "Understanding the water crisis in Africa and the Middle East: How can science inform policy and practice?" *Bulletin of the Atomic Scientists* 71(2): 39–49.

Jackson, E. T., and S. Gariba. 2002. "Complexity in local stakeholder coordination: Decentralization and community water management in northern Ghana." *Public Administration and Development: The International Journal of Management Research and Practice* 22(2): 135–140.

Jami, A. A., and P. R. Walsh. 2017. "From consultation to collaboration: A participatory framework for positive community engagement with wind energy projects in Ontario, Canada." *Energy research & social science* 27: 14–24.

JISC. "A five step risk management model." https://www.jisc.ac.uk/guides/risk-management/five-step-model.

Jimu, I. M. 2008. "The role of stakeholders in the provision and management of water kiosks in Nkolokoti, Blantyre (Malawi)." *Physics and Chemistry of the Earth, Parts A/B/C* 33(8–13): 833–840.

Keega, M. 2017. "Institutionalizing WASH capacity development in South Sudan: moving from emergency response to development."

Kendie, S. B. 1992. "Survey of water use behavior in rural North Ghana." *Natural resources forum* 16(2): 126–131.

Kleemeier, E. 2000. "The Impact of Participation on Sustainability: An Analysis of the Malawi Rural Piped Scheme Program." *World Development* 28(5): 929–944.

Komives, K., B. Akanbang, R. Thorsten, B. Tuffuor, W. Wakeman, E. Larbi, and D. Whittington. 2008. "Post-construction support and the sustainability of rural water projects in Ghana."

Kuper, M., M. Dionnet, A. Hammani, Y. Bekkar, P. Garin, and B. Bluemling. 2009. "Supporting the shift from state water to community water: lessons from a social learning approach to designing joint irrigation projects in Morocco." *Ecology and society* 14(1): 19.

Makawy, A. Y. I. 2013. "Transboundary Water in Sudan Post the Separation of South Sudan." Faculty of Engineering, University of Khartoum.

Marks, S. J., K. Komives, and J. Davis. 2014. "Community participation and water supply sustainability: evidence from handpump projects in rural Ghana." *Journal of Planning Education and Research* 34(3): 276–286.

Matamula, S. 2008. "Community-based management for sustainable water supply in Malawi."

Matoso, M. 2018. "Supporting sustainable water service delivery in a protracted crisis: Professionalizing community-led systems in South Sudan."

McConville, J. R., and J. R. Mihelcic. 2007. "Adapting life-cycle thinking tools to evaluate project sustainability in international water and sanitation development work." *Environmental Engineering Science* 24(7): 937–948.

McNiff, J. 2013. *Action Research: Principles and practice*, 3rd ed. Routledge.

Mimrose, D. M. C. S., E. R. N. Gunawardena, and H. B. Nayakakorala. 2011. "Assessment of sustainability of community water supply projects in Kandy District."

Moon, S. 2006. "Private operation in the rural water supply in central Tanzania: Quick fixes and slow transitions." *WaterAid Tanzania*.

Morris, L., M. Ma, and P. Wu. 2014. *Agile Innovation: The Revolutionary Approach to Accelerate Success, Inspire Engagement, and Ignite Creativity*. John Wiley & Sons.

Morse, J. M. 2016. *Mixed-Method Design: Principles and Procedures*. Routledge.

Mtinda, E. O. 2006. "Sustainability of Rural Water supply and sanitation services under community management approach: The case of six villages in Tanzania." Department of water and environmental study, Linkopings Universiet.

Muriana, C., and G. Vizzini. 2017. "Project risk management: A deterministic quantitative technique for assessment and mitigation." *International Journal of Project Management* 35(3): 320–340.

Mvulirwenande, S., U. Wehn, and G. Alaerts. 2017. "Evaluating knowledge and capacity development in the water sector: Challenges and progress." *Water International* 42(4): 372–384.

Nkongo, D., and W. Tanzania. 2009. "Management and regulation for sustainable water supply schemes in rural communities." *WaterAid Tanzania*.

Nyakwaka, S., and M. K. Benard. 2019. "Factors influencing sustainability of community-operated water projects in central Nyakach Sub-County, Kisumu County, Kenya." *International Journal of Academic Research in Business and Social Sciences* 9(7). https://doi.org/10.6007/ijarbss/v9-i7/6096.

Nzengya, D. M. 2015. "Exploring the challenges and opportunities for master operators and water kiosks under Delegated Management Model (DMM): A study in Lake Victoria region, Kenya." *Cities* 46: 35–43.

Oino, P. G., G. Towett, K. K. Kirui, and C. Luvega. 2015. "The dilemma in sustainability of community-based projects in Kenya." *Global journal of advanced research* 2(4): 757–768.

Olechowski, A., J. Oehmen, W. Seering, and M. Ben-Daya. 2016. "The professionalization of risk management: What role can the ISO 31000 risk management principles play?" *International Journal of Project Management* 34(8): 1568–1578.

Omar, Y. Y., A. Parker, J. A. Smith, and S. J. Pollard. 2017. "Risk management for drinking water safety in low and middle income countries - cultural influences on water safety plan

(WSP) implementation in urban water utilities." *Science of The Total Environment* 576: 895–906. https://doi.org/10.1016/j.scitotenv.2016.10.131.

Ong'wen, O. 1996. "NGO experience, intervention, and challenges in water strain, demand, and supply management in Africa." *Water Management in Africa and the Middle East: Challenges and Opportunities*: 274.

Oshionebo, E. 2019. "Community Development Agreements as Tools for Local Participation in Natural Resource Projects in Africa." *Human Rights in the Extractive Industries*: 77–109.

Oyebande, L. 2001. "Water problems in Africa—how can the sciences help?" *Hydrological Sciences Journal* 46(6): 947–962.

Patton, M. 2002. *Qualitative Research and Evaluation Methods*, 3rd ed. Thousand Oaks, Calif.: Sage.

Pink, M. A., Y. Taouk, S. Guinea, K. Bunch, K. Flowers, and K. Nightingale. 2016. "Developing a conceptual framework for student learning during international community engagement." *Journal of University Teaching & Learning Practice* 13(5): 21.

Piperopoulos, P. 2012. *Entrepreneurship, Innovation and Business Clusters*. Taylor and Francis.

Qazi, A., J. Quigley, A. Dickson, and K. Kirytopoulos. 2016. "Project Complexity and Risk Management (ProCRiM): Towards modelling project complexity driven risk paths in construction projects." *International Journal of Project Management* 34(7): 1183–1198.

Rae, D. 2007. *Entrepreneurship: From Opportunity to Action*. Palgrave Macmillan.

Rathgeber, E. 1996. "Women, men, and water-resource management in Africa." *Water management in Africa and the Middle East: Challenges and opportunities*: 49–69.

Richmond, A. K. 2019. "Water, Land, and Governance: Environmental Security in Dense Urban Areas in Sub-Saharan Africa." *The Environment-Conflict Nexus*: 91–102.

Rutten, M., T. Dietz, D. Foeken, G. Seuren, and F. Veldkamp. 2014. "Water dynamics in the seven African countries of Dutch policy focus: Benin, Ghana, Kenya, Mali, Mozambique, Rwanda, South Sudan."

Sadgrove, K. 2016. *The complete guide to business risk management*. Routledge.

Salman, S. M. 2011. "The new state of South Sudan and the hydro-politics of the Nile Basin." *Water International* 36(2): 154–166.

———. 2014. "Water resources in the Sudan North-South peace process and the ramifications of the secession of South Sudan." *Water and post-conflict peace building*.

Saunders, M., P. Lewis, and A. Thornhill. 2016. *Research methods for business students*, 7th ed. England: Pearson Education, Harlow.

Siemens Stifrung. "Safe Water Enterprises." https://www.siemensstiftung.org/en/projects/safe-water-enterprises.

Silvestri, G., J. Wittmayer, K. Schipper, R. Kulabako, S. Oduro-Kwarteng, P. Nyenje, and R. van Raak. 2018. "Transition management for improving the sustainability of WASH services in informal settlements in Sub-Saharan Africa—an exploration." *Sustainability* 10 (11): 4052.

Smith, P. 2013. "Environmental Health." *Reference Module in Earth Systems and environmental sciences.* https://doi.org/10.1016/B978-0-12-409548-9.05919-4.

Stoyanova, Z., I. Petkova, and K. Todorova. 2018. "Risk Management Strategies in Water Projects in Bulgaria." *Water Projects in Bulg* (2): 228–238. https://www.unwe.bg/uploads/Alternatives/Zornica_EAlternativi_2_2018-6.pdf.

Sullivan, C. 2002. "Calculating a water poverty index." *World Development* 30(7): 1195–1210.

Swain, A. 2002. "The Nile River Basin Initiative: too many cooks, too little broth." *SAIS Review* 22(2): 293–308.

Therkildsen, O. 1988. "Watering white elephants? Lessons from donor-funded planning and implementation of rural water supplies in Tanzania."

Thirst Project. "Standards for Implementing Water Projects." Retrieved December 21,2019. https://www.thirstproject.org/wp-content/uploads/2015/05/standards-for-sustainability.pdf.

UN-GLAAS. "UN-water global annual assessment of sanitation and drinking-water (GLAAS) 2012 report: the challenge of extending and sustaining services." https://apps.who.int/iris/handle/10665/44849.

UNICEF. "Water, Sanitation, and Hygience South Sudan Briefing Note." Retrieved December 17, 2019. https://www.unicef.org/southsudan/media/2076/file/ percent20UNICEF-South-Sudan-WASH-Briefing-Note-Oct-2019.pdf.

———. "Programming for Sustainability in Water services- A Framework." Retrieved December 21, 2019. https://www.unicef.org/wash/files/Programming_for_Sustainability_in_Water_Services_A_Framework.pdf.

USAID. "Draft Water, Sanitation and Hygiene (WASH) Program 2013-2018." Retrieved December 20, 2019. https://www.usaid.gov/sites/default/files/documents/1860/USAID percent20South percent20Sudan percent20Draft percent20Water, percent20Sanitation percent20and percent20Hygiene percent20Program percent202013-2018.pdf.

Vijaya, S. M. 2016. "Lean six sigma project management – a stakeholder management perspective." *TQM Journal* 28(1): 132–150. https://doi.org/10.1108/TQM-09- 2014-0070.

Von Korff, Y., K. A. Daniell, S. Moellenkamp, P. Bots, and R. M. Bijlsma. 2012. "Implementing participatory water management: recent advances in theory, practice, and evaluation." *Ecology and Society* 17(1).

Walliman, N. 2011. *Your Research Project: designing and planning your work*, 3rd ed. Sage Publications.

Wendl, A. K. 2016. "International water rights on the White Nile of the new state of South Sudan." *BC Int'l & Comp. L. Rev.* 39: 1.

Whaley, L., and F. Cleaver. 2017. "Can 'functionality' save the community management model of rural water supply?" *Water resources and rural development* 9: 56–66.

Whittington, D., J. Davis, L. Prokopy, K. Komives, R. Thorsten, H. Lukacs, and W. Wakeman. 2009. "How well is the demand-driven, community management model for rural water supply systems doing? Evidence from Bolivia, Peru, and Ghana." *Water Policy* 11(6): 696–718.

Wickham, P. A. 2006. *Strategic Entrepreneurship*, 4th ed. Prentice Hall.

Wiles, R. 2012. *What are qualitative research ethics.* London: Bloomsbury.

Zuber-Skerritt, O. 2015. "Participatory action learning and action research (PALAR) for community engagement: A theoretical framework." *Educational research for social change* 4(1): 5–25.

List of Appendices

Appendix 1: Summary of Interviews Conducted

Intvw #	Names in Full	Organization	Position	Interview Length
1.	Hon. Gbamisi Charles Babiro	County Government	Former county commissioner	37.05 minutes
2.	Mrs. Grace Amjuma	State Government	Civil servant	21.17 minutes
3.	Mr. Edward Dazangapai	Businessman	Water point manager	16.33 minutes
4.	Mr. Samuel Gume	Yubu Development Agency (YDA)	Water point manager	21.20 minutes
5.	Ms. Lina Leone	Private business	Water point manager	22.18 minutes
6.	Dr, Duboul	Tambura Civil Hospital	Medical doctor	24.11 minutes
7.	Mr. Marko Adidi	Businessman	Water technician	26.11 minutes
8.	Mr. Khamis Nando	Water Management Association	Water point manager	23.26 minutes
9.	Mr. Simon Kukuk Kumborani	Deputy Secretary General	Water user	8.32 minutes
10.	Mr. James Mopaya	Director of Resolution	Water user	15.35 minutes
11.	Mr. Michael Zungua	Businessman	Kiosk supervisor	12.13 minutes
12.	Mr. Michael Sababa	Chief	Water point manager	16.17 minutes
13.	Mr. Dungura	Executive Director	Stakeholder	26.08 minutes
14.	Mr. John Louis Barikpio	Inspector of Cooperative	Water point manager	13.18 minutes

15.	Mr. Nafuoni Tartizio	Director of Account	Community elder	18.32 minutes
16.	Mr. Mitedio	Water technician	Water supervisor	25.02 minutes
17.	Mr. Henry Zinaro	Payam Administrator	Water point manager	27.52 minutes
18.	Mr. Peter Nazaro	Elder in the community	Water point manager	23.23 minutes
19.	Elda Martha Bayu	GIZ	Coordinator	24.48 minutes
20.	Hon. Derick Sasa	Government	Commissioner	15.35 minutes
				416.56 minutes
				6.94 hours

Appendix 2: Letter of Introduction from LIGS University

LIGS University
810 Richards Street, Suite 836
Honolulu
HI 96813 USA

25th September 2020

Dear Sir/Madam,

Re: Recommendation for Tamburo Michael Renzi to conduct a Doctor of Philosophy (Project Management) research

This letter confirms that above named person is a student of the Ph.D. program at LIGS university and is currently conducting a research as part of his compulsory study activities.

The extensive research work will lead to a thesis comprising 50,000 words. The student's selected topic is 'Analysis of Risk Management of Water Kiosks Projects In Africa' A Case Study Of South Sudan.

Kindly grant this student the required permit.

Sincerely,

Tereza Kubankova, Registrar
Office of Admissions & Records
Tereza.kubankova@ligsuniversity.com

Appendix 3: List of Questionnaires

LIGS University USA
Research Questionnaires for Project Managers

Questionnaire #:

I want to thank you for taking the time to meet with me today. My name is Tamburo Michael Renzi. I would like to talk to you regarding your experiences in participating community water kiosk projects. My PhD study is about risk management in community water kiosk projects in South Sudan rural areas. The interview should take about forty-five minutes. All responses will be kept confidential and used for research purposes only. It will not be possible to identify any particular individual or address in the results.

Before we proceed, do you have any question about what I have just explained?

_____

_____

1. Personal details:

1.1. Name: _____ (optional)
1.2. Gender: 1. Male ☐   2. Female ☐
1.3. Occupation: _____
1.4. Education:

1. Ph.D.
2. Master's Degree
3. Bachelor's Degree
4. Diploma
5. Senior Secondary School

1.5. Years of Experience
   a. 1–5 years
   b. 5–10 years
   c. 10–15 years
   d. 15–20 years
   e. Above 20 years

1.6 Age Group
   a. 20–30 years
   b. 30–40 years
   c. 40–50 years
   d. 50–60 years
   e. Above 60 years

2. Have you ever directly managed a community water project? *Please tick the box*

   1. Yes ☐   2. No ☐

3. If yes, kindly state the number of the projects and their duration in months

   _____
   _____

4. Do you think the project planning through implementation was done the right way?

       1. Yes ☐   2. No ☐

5. If No, kindly elaborate where you think the implementers went wrong.

   _____
   _____

6. At what point did you face challenges in the project implementation? ***Please choose only one from the list.***

   1. Site identification
   2. Construction water kiosk
   3. Operation of water kiosk
   4. Training and handover of the facility
   5. Monitoring and Evaluation
   6. Others, please state
      a. _____
      b. _____

7. What challenges did you face while managing these projects?

   1. Cultural issues
   2. Communication
   3. Financial
   4. Logistics
   5. Ownership
   6. Others, please state
      a. _____
      b. _____

8. Do you think that the project/s is/are exposed to any type of risk?

   1. Yes ☐   2. No ☐

9. If yes, what type of risks? Please select from the list below:

   1. Legal risk
   2. Political risk
   3. Financial risk
   4. Technical risk
   5. Security risk
   6. Management/administrative risks
   7. Others, please state
      a. _____
      b. _____

10. Could you kindly state some possible solutions to the above risks?

_____

_____

_____

11. Are there some guidelines/policies in place to guide project implementation by the community?

    1. Yes ☐   2. No ☐

12. If yes, please elaborate them?

_____

_____

_____

13. Do you think this project will exist in the next ten years?

    1. Yes ☐   2. No ☐

14. If no, kindly explain the reasons

_____

_____

_____

15. What are some of the solutions that can used to extend water project sustainability?

   a. Harmonize policies
   b. Capacity training
   c. Reorganize water rates
   d. Control number of customers
   e. Regular maintenance
   f. Control fetching time
   g. Periodical monitoring of facility

16. What recommendations do you have towards the improvement of the water project?

   1. Stakeholders involvement
   2. Regular meetings
   3. Timely supervision
   4. Alternative funding
   5. Others, please state
      a. _____
      b. _____

LIGS University USA

Research Questionnaires for Water Vendor

I want to thank you for taking the time to meet with me today. My name is Tamburo Michael Renzi. I would like to talk to you regarding your experiences in participating community water Kiosks projects. My PhD study is about risk management in community water kiosk projects in South Sudan rural areas. The Interview should take about thirty minutes. All responses will be kept confidential and used for research purposes only. It will not be possible to identify any particular individual or address in the results.

Before we proceed, do you have any questions about what I have just explained?

_____

_____

1. Personal details:

1.1. Name: _____ (optional)

1.2. Gender: 1. Male ☐   2. Female ☐

1.3. Occupation: _____

1.4. Education:
   1. Ph.D.
   2. Master's Degree
   3. Bachelor's Degree
   4. Diploma

5. Senior Secondary School

1.5. Years of Experience

    a. 1–5 years

    b. 5–10 years

    c. 10–15 years

    d. 15–20 years

    e. Above 20 years

1.6. Age Group

    a. 20–30 years

    b. 30–40 years

    c. 40–50 years

    d. 50–60 years

    e. Above 60 years

1.7. Are you a resident of this village? (Please tick the box)

    1. Yes ☐   2. No ☐

1.8. How many people live in your house? Please choose from the following options:

    1. 1–5

    2. 6–10

    3. 11–15

    4. 16–20

    5. Above 20

Water consumption

2. What is the main source of water in this village? *Please choose only one from the options.*

   1. Tap water (community water Kiosk) ☐
   2. Rivers ☐
   3. Natural spring ☐
   4. Rain water ☐
   5. Borehole (open well) ☐

3. For how long has the water project served the community members? *Please choose from the followings:*

   a. 1–5 Years
   b. 5–10 years
   c. 10–15 years
   d. 15–20 years
   e. Above 20 years

4. Which member of your household fetches water in the kiosk (water tape)? Choose from the following options. ***You can select more than one.***

   1. Mother
   2. Father
   3. Children (a) Girls (b) Boys
   4. House Help (a) Female (b) Male

5. Are there times of the day when you feel overwhelmed with work?

      1. Yes ☐   2. No ☐

6. If yes, choose from the following:

    1. Morning Hours (6:00–11:00 a.m.)
    2. Afternoon Hours (12:00–4:00 p.m.)
    3. Evening Hours (5:00–9:00 p.m.)

7. Do you think that the water is safe for drinking?

    (A). 1. Yes ☐   2. No ☐

    (B). If no, what makes the water not safe for drinking? *Select from the following options.*

    1. Not treated
    2. Contaminated
    3. Polluted
    4. Unclean environment
    5. Pipes/water tanks are broken

8. What are the most common uses of water fetched in this water kiosk/tape water? **Please choose only one.**

    1. Domestic use ☐
    2. Construction ☐

3. Irrigation ☐
4. Cattle/Livestock ☐
5. Other uses _____ (specify)

9. How do the communities perceive this water kiosk project? **Please choose only one.**

   1. Very Important
   2. Important
   3. Not important
   4. Very unimportant

10. What do you think are some of the challenges facing this water project? *Please choose those that apply.*

    1. Lack of capacity
    2. Lack of regulations
    3. Reluctant to pay
    4. High demand
    5. Unsafe water
    6. Poor management
    7. Theft of equipment

11. What solutions do you propose to solve these challenges? *Please choose those that apply.*

    1. Capacity trainings
    2. Increase water production
    3. Introduce regulations
    4. Treating the water
    5. Redesign payment systems
    6. Provision of security
    7. Others
        a. _____
        b. _____

12. Are there any risks associated with this water project?

    (A). 1. Yes ☐   2. No ☐

    (B). If yes, which of the following describe the risks associated with the water kiosk project? Please choose those that apply.

    1. Conflict of interest
    2. Breakdown of equipment
    3. Cultural perception
    4. Financial issues

5. Administrative issues
6. Theft of equipment
7. Others
   1. _____
   2. _____
   3. _____

www.ingramcontent.com/pod-product-compliance
Lightning Source LLC
Chambersburg PA
CBHW020632220526
45464CB00001B/114